Time *to* Chang

Rethinking
the Big Bang Theory
and Black Holes

Unifying gravity, Relativity and the
quantum mechanics of all universes

Roger Wood

ISBN: 979-8-78-405159-2

Author Biography

Dr. Roger Wood is a Senior Lecturer in Education at Oxford Brookes University, having previously been a Senior Lecturer at Bishop Grosseteste University and a Lecturer (Research) in Education at the University of Aberdeen. Prior to his university teaching and research career, Dr. Wood was a headteacher and teacher for 23 years.

Dr. Wood's philosophy-informed research focuses upon interpreting and understanding outcomes of human philosophizing, including confidence-informed social motivation and the impact upon the context-based endeavours of children, adolescents and teachers. This includes identifying teacher behaviours and methods that enhance students' motivation for and engagement with school-based formal and informal learning.

Dr. Wood is a scientist and Chartered Biologist by background, specialising in marine biology, primatology and cosmology. His work in science education, conservation, primatology and cosmology has led to his election to a number of Fellowships including the Royal Astronomical Society, the Royal Society of Biology, the Royal Scottish Society of Arts, and the Linnean Society. In addition, he is a Fellow of the Higher Education Academy and a Fully Registered Teacher with the General Teaching Council for Scotland.

Dr. Wood's education research focuses upon strategies for enhancing children's engagement with Science and STEM through University-initiated approaches to supporting the career-long professional learning and confidence of primary science teachers. This places an emphasis upon enhancing our knowledge and understanding of the factors and strategies that influence children's engagement with primary science education (including STEM). Impacts include the development of university and school-based strategies that enhance trainee and qualified teachers' confidence when embedding inquiry-based learning within their own classrooms.

Acknowledgements

The further development of GLEW Theory and its evolution into UE-GLEW Theory was the result of innumerable enjoyable hours in the University of Cambridge Library and the Radcliffe Science Library in Oxford. Gaining access to the resources that I needed was due to the extremely helpful librarians within both libraries. Through their invaluable help, much of my research, writing and editing has been carried out in blissful academic isolation and rumination, alongside discussions with many insightful and imaginative individuals.

Thank you, Oscar, for our discussions regarding there always having to be something and there never being nothing!

As always, my family have remained as the source of my inspiration, pride, love, fulfilment and contentment throughout this path through my thought experiments and learning journey: Ingrid, David, Mike, Jenny, Khloe, Erin, Logan, Oscar, Callum and James.

This book is dedicated to

Charlie

Affectionately known as **Slow Thinker's Cat**

Always one to look to the stars and make his opinions known, as long as there was time set aside for food and a snooze in the laundry basket. (Charlie that is, not me!)

So what do you think of our latest theory?!

My work here is done!

4th October 2021

CONTENTS

Preface

Albert Einstein, in one of his many oft-quoted inspirational moments, was heard to say, "I believe in intuitions and inspirations. I sometimes feel that I am right. I do not know that I am". However, despite being attributed to an even more famous quote, Einstein did not state that, "Insanity is doing the same thing over and over and expecting different results". Both quotes have underpinned my thinking and the development of the cosmological and gravitational theory discussed within this book.

This new volume centres upon several fundamental ideas and changes that have emerged since the publication of *The new Big Bang Theory, Black Holes and the Multiverse explained (Gravity-Light Energized Waves as the GLEW holding the Multiverse together: rethinking the composition and function of black holes, dark energy and dark matter)* in May 2020.

By writing this book, informed by *The new Big Bang Theory, Black Holes and the Multiverse explained*, I fully appreciate that what I am postulating flies in the face of over 100 years of gravitational, quantum and cosmological theory. However, by doing so, I am presenting new approaches to thinking that will, hopefully, lead us to realise that not only does our thinking about gravity and light need revision but also that we can begin to make great sense of how gravity light, and black holes interact. Within this book, new thinking and concepts are suggested for discussion and debate by scientists and the general public alike. These are collated as UE-GLEW theory. This novel theory centres upon Universal Energy (UE) as the basis of all matter, energy, momenta and gravitational effects within our Universe. UE will never be directly observed at the quantum level but the impact of UE can be imagined and interpreted based upon all that we are able to observe. There are several approaches that we can take to thinking about the quantum components of the universe, if we are to understand their mechanical activities and consequences. As with different states of matter on Earth, dark matter exists in several different states, or forms, simultaneously. These different states are dark matter, gravity, light and dark energy. All four are simultaneously known as Universal Energy.

The key question that is addressed by UE-GLEW theory is "How does the Universe work?" (I have referred specifically to the Universe that we occupy, and are transient guests within, through the use of a capital letter. Other universes within the Multiverse, have referred to with a lower case 'u').

This book is the natural evolution of the ideas discussed in *The new Big Bang Theory, Black Holes and the Multiverse explained (Gravity-Light Energized Waves as the GLEW holding the Multiverse together: rethinking the composition and function of black holes, dark energy and dark matter)*. This evolution incorporates some elements of the original GLEW and GLARE theory, whilst adding a number of new theoretical ideas regarding gravity, spacetime geometrics, the theory of general relativity, and the form and function of black holes as thermodynamic regulatory gateways (TRGs). Chapter 3 discusses the explicit links between GLEW Theory and the key theories that have led to the ideas central to GLEW.

There is always something. Nothing does not exist! Where we believe there to be nothing, there will always be something. There has always been something, even in situations where we believe there may once have been nothing. Between us and the clouds, and across the atmosphere, there is air: we cannot always see it and the atoms that comprise such air but we know that it is there. There is something, even though it is usually invisible or undetectable.

The same can be said of what lies between the planets, stars and other objects within the Universe and the many universes that comprise the Multiverse. Our mind and senses are limited in our ability to detect, perceive and interpret. Equally, the vast majority of the Earth's human population are limited in our ability to understand the science and invent the technology needed to conclusively detect even fundamentals such as dark matter, dark energy, and black holes. We know that these exist on the basis of the detectable impact and actions. However, as Professor Stephen Hawking (1942 – 2018) once stated: "… it is very difficult to make a mark for oneself in an experimental subject. One is often only part of a large team, doing an experiment that takes years. On the other hand, a theorist can have an idea in a single afternoon, or, in my case, while getting into bed, and write a paper on one's own or with one or two colleagues to make one's name".

To understand cosmology and gravitation, in particular, we need theories that are embedded within the thought boundaries of the human intellect, and that may be understood and, wherever possible, applied. In the writing of this second edition of the GLEW and GLARE theory, I have drawn upon a wide range of sources. These are listed at the end of this book. The work of Albert Einstein, Richard Feynman and Wolfgang Pauli, as well as Stephen

Hawking and Roger Penrose, have been extensively consulted. The 2019 Nobel Prize for Physics 2019 rewarded new understanding of the Universe's structure and history, and the first discovery of a planet orbiting a solar-type star outside the solar system. The 2019 Nobel Laureates, including Professor Roger Penrose, were recognised for their contributions to answering fundamental questions about our existence. These include ' What happened in the early infancy of the universe and what happened next?', and 'Could there be other planets out there, orbiting other stars'?

The Universal Energy: Gravity-Light Energized Waves (UE-GLEW) theory, presented and discussed within this book, is intended as another contribution to informing and clarifying our understanding. It is acknowledged that, when trying to understand the governing dynamics of the Multiverse and the Universe, we may be perpetually searching for the 'needle in the haystack. Therefore, when considering the theory herein, and, indeed, any other cosmological gravitational theory, if we are to go searching for the needle in the haystack, we need to be mindful of four things:

1. To be sure that the needle or a needle is actually *in* the haystack.
2. To be clear why it is important to find the needle.
3. To be aware of whether there might be something to find that is more important than the needle!
4. Being willing to rethink what the needle might be like, in terms of size, form and function.

Roger Wood

Perthshire, Oxford and Falmouth

Completed in Falmouth
December 2021

CHAPTER 1

Universal Energy as Gravity-Light Energized Waves leading to SpaceFabric Curvature

The UE-GLEW Theory

Introduction: the 'big' unsolved problems within cosmological physics

Cosmology is the study of the largest scales in the universe, such as the longest times and the furthest distances (both of which are, in human terms, inter-related). There are a huge number of unsolved problems, which require an understanding of fundamental physics and its relationship to data (see Brooks, 2005).

Some of these unsolved problems, as listed below, have been addressed and redefined through the lens of this latest evolution of the UE-GLEW theory (Universal Energy: Gravity-Light Energized Waves). These include:

1. The problem of time, where, in quantum mechanics, time is regarded as a classical background parameter, with the flow of time being potentially universal and absolute. However, in general relativity time is the fourth component of four-dimensional spacetime: the flow of time changes depending on the curvature of spacetime and the spacetime trajectory of the observer. The puzzle that faces physicists and cosmologists is how the two concepts of time can be reconciled or aligned.

2. The Horizon problem: why the distant universe appears to be homogeneous when the Big Bang Theory is predictive of larger measurable anisotropies than those which, to date, have been observed. Cosmological inflation has been generally accepted as the solution, but it may be that there are other possible explanations such as a variable speed of light (or VSL) (Fixsen, 2009; Remmen and Carroll, 2014).

3. The origin and future of the universe that we inhabit. That is, how was the universe created, and how will it, if at all, come to an end? It may be that there is no end to the universe but, instead, that it is part of an infinitely recurring cyclic model.

4. The identity and composition of dark matter (DM). Puzzles include its structure, possibly as a particle, or as the basis of phenomena that is an extension of gravity.

5. Dark energy (DE) may be the cause of the accelerated expansion of the universe: this is also known as the de Sitter phase. A key puzzle relates to why the energy density of the dark energy component of the same magnitude as the density of matter at present when the two evolve quite differently. One solution may be that dark energy is a cosmological constant.

6. Dark flow has also been mooted, as a non-spherically symmetric gravitational pull from outside the observable universe. This may be responsible for some of the observed motion of large objects such as galactic clusters in the universe.

7. In terms of gravitational origins and effects, this leads to the puzzles surrounding the nature of quantum gravity. Specific puzzles include the question of whether quantum mechanics and general relativity can ever be fully realised as a consistent theory, such as a quantum field theory. This depends upon whether spacetime fundamentally continuous or discrete. In addition, a consistent theory might involve a force mediated by gravitons, as discrete particles, as gravity may be the product of a discrete structure of spacetime itself, as with loop quantum gravity. Finally, deviations from the predictions of general relativity may be possible at very small or very large scales or in other extreme circumstances as the result of a quantum gravity mechanism.

Further to the seven problems, or puzzles, above, we can then begin to consider puzzles relating to black holes. These puzzles include, but are not limited to, Hawking black hole radiation and the black hole information paradox. Further puzzles lead us to consider whether black holes actually produce thermal radiation, as expected on theoretical grounds, and whether such radiation contains information about their inner structure. Gauge-gravity duality suggests that black holes reveal information about their inner structure, whilst Hawking's original calculation suggests that they do not.

Additional puzzles and ideas include supermassive black holes as the potential origins of galaxy velocity dispersion, and large-scale anisotropy where the universe is anisotropic, thus making the cosmological principle invalid. Of particular note is the data from the NRAO VLA Sky Survey (NVSS) catalogue, which is inconsistent with the local motion as derived from Cosmic Microwave Background Radiation (CMBR). This, alongside other NVSS radio data, reveals large-scale anisotropy. Indeed, the cosmic-microwave-background data shows several features of anisotropy, which are not consistent with the Big Bang model (see Website Links at the end of this book).

In addition, there are puzzles surrounding the verifiable percentages of dark matter (DM) and dark energy within the universe. Based upon various data surveys, researchers have, in the main, stated that 31.5% of the matter-energy density of the universe consists of DM and Ordinary Matter (OM). For example, NASA have asserted that dark matter makes up approximately 27% of the universe, with the other 68% and 5% being dark energy and 'normal matter' (observable objects) respectively. Recent figures, confirmed in October 2020, further to research from the University of California, in Riverside, and the National Research Institute of Astronomy and Geophysics in Egypt, suggested that dark matter is essentially 21.6% of the universe. This differs from NASA, who have estimated that dark matter is equal to 24% of the Universe with another 4.6% being ordinary matter (atoms), leaving 71.4% as dark energy (DE). Other published estimates posit DM as 26.8%, OM as 4.9% and DE as 68.3%. Therefore, the matter-energy density is only approximate as there is variance in DM between 24% and 27%, whilst DE varies, according to figures, between 68% and 71% (see Website Links for sources).

It appears that humans carry a naïve, and, possibly, egotistical assumption that the secrets (or puzzles) of the universe will be revealed to them, and, that it will be possible to interpret these secrets to the point of comprehension and understanding them in the form of definitive knowledge. Scientific and philosophical thinking, both of which are interlinked, are, to paraphrase Einstein, best unleashed through imagination and creative thinking bordering on daydreams. Theoretical thinking requires rich imagination. Such imaginative thinking is most effectively undertaken by uncommon and individual minds, that are unfettered and unconstrained by dogma imposed by authority. For such individuals, speculation is a passion.

We need, therefore, to approach knowledge and understanding as puzzles for speculation and investigation, rather than as problems.

Much of the acclaim that 'celebrity' scientists have received has been as much for their media-projected image, as it has been for the incomprehensibility of their theories by the majority. Many people with a non-scientific background have bought their books and read their public-focused publications but such books are soon set aside, by their intended audience, as the ideas are presented in such a way as to be unfathomable. Therefore, it is vital that any theory, whether of the mechanics which govern the dynamics of the universe or a unified field theory, is comprehensible and understandable. This should include elements of common sense thinking and ideas, that appeal to what is thinkable by the majority.

A positive means of inspiring theoretical thinking that is rich in imagination, creativity, and original thinking is central to developing possible solutions to the big puzzles within physics. Therefore, we need to approach human knowledge and understanding of cosmology as puzzles that provide a basis for speculation and investigation. This is a far more positive approach than perceiving prior cosmological conundrums as problems. Through such puzzles and associated free speculation, we can begin to rethink how we approach several key ideas and definitions relating to the mechanics and dynamics of the universe.

The unsolved puzzles within physics, lie not with either physics or the processes themselves but with human awareness, knowledge and interpretations. There are five key barriers to human knowledge, understanding and endeavour. These are:

1. The epistemic foundations of our current understanding, alongside;
2. Our ability to formulate and make sense of hermeneutic interpretations.
3. We are further confounded by the extent of our imagination, and;
4. Willingness to accept new ideas and thinking as plausible.
5. Finally, language can be limiting in terms of the ability articulate and describe phenomena.

Through approaching our thinking through other lenses, as means of interpreting our ideas and hypotheses in a different way, we are more likely to be able to understand some of the key concepts relating to the universe that, currently, remain unsolved due to the epistemic foundations and interpretive means we have been insisting upon using for almost four hundred years!

On 27th May 2021, the BBC announced that a new dark matter map had revealed a cosmic mystery. The Dark Matter map was developed by the Dark Energy Survey Collaboration, led by Dr. Niall Jeffrey, of École Normale Supérieure, in Paris [see Website link 1]. On the DM map, the black areas are vast areas of 'nothingness', called voids, where the laws of physics might be different. The bright areas are where dark matter is concentrated. These are called "halos", in the midst of which are the galaxies. The map suggests that galaxies are part of a larger invisible structure. However, this new dark matter map is not showing quite what astronomers expected. The results of the survey, of 100 million galaxies, suggest that dark matter is slightly smoother and more spread out than the current best theories predict. The observation appears, in particular, to stray from Einstein's theory of general relativity, thereby posing a conundrum for researchers. As Dr. Jeffrey informed BBC News, "If this disparity is true then maybe Einstein was wrong …You might think that this is a bad thing, that maybe physics is broken. But to a physicist, it is extremely exciting. It means that we can find out something new about the way the Universe really is" [Website link 1].

The DM map, based upon DES data, has evolved from previously reported findings in 2015 and 2017. In 2015, dark matter was detectable through light distortions, caused by dark matter's gravitational "lensing" of passing detectable light. The ultimate aim of the DES project is to test the idea of dark energy. The expansion of the universe is happening at an increasing rate, and dark energy has been suggested as the cause of the acceleration. In addition, it was stated that it is possible, with the emergent data for DM and DE, that dark energy may not be a good theory for explaining why the universe behaves as it does, and in fact, general relativity itself is wrong [Website link 5]. Two years later, in 2017, Professor Ofer Lahav of University College London (UCL), and chair of the DES Advisory Board, unveiled the second DM map, having surveyed 26 million galaxies. (For context, 100 million galaxies have been surveyed to create the 2021 DM map). Lahav stated that, "Dark energy and dark matter represent probably one of the biggest mysteries in the world of science …this has generated a lot of interest across the whole of science because it is a major shakeup … actually we still don't know what it is," he said. Scientists have suspected that there is more material in the Universe than we have been able to observe in the past 80 years. The movement of stars and galaxies indicate that the Universe, in addition to dark energy (DE) is also made up of invisible particles called dark matter [Website link 6].

The DM maps of 2021, 2017 and 2015 have enhanced our understanding of the first DM map, which was published in 2007. Data from the Hubble Space Telescope provided the best evidence, at that time, that the distribution of galaxies follows the distribution of dark matter. However, it that point, it was announced that "…dark matter does not reflect or emit detectable light, yet it accounts for most of the mass in the Universe" [see Website link 4]. This, according to Massey et al., (2007), appears to be due to dark matter attracting ordinary matter through its gravitational pull. To be detectable on Earth, the light from galaxies has to pass through intervening dark matter. This dark material bends light in much the same way as light is bent when travelling through a lens. According to Dr. Richard Massey, from the California Institute of Technology (Caltech), "We understand statistically what those galaxies are supposed to look like. If you place some dark matter in the way, this dark matter - through its gravity - bends the path of light … As the light gets deflected, it distorts the shape of the background galaxies. So we end up seeing them in a distorted way, as if through lots of little lenses - and each of those lenses is a bit of dark matter." Interestingly, the 2007 results indicated that the Universe is made up of 4% ordinary matter, 26% dark matter and 70% dark energy.

The idea of DE is comparatively new in relation to some of the longer-held ideas and theories within cosmology and physics. In 1998, two teams of astronomers discovered that the expansion of the Universe was accelerating, rather than slowing down, as the theory at the time suggested. Physicists speculated that the acceleration was caused by something they named 'dark energy' (abbreviated herein as DM).

Reflecting upon the findings presented within the 2021 DM map, Professor Carlos Frenk, of Durham University, stated that cosmologists, astrophysicists may have to have to look at the possibility of new physics to explain what has been observed. This would be necessary as, until then, there are no solid grounds to explore our knowledge and understanding, because there is, at present, no theory of physics to guide scientists' interpretations. Furthermore, Professor Lahav, of the DES and lead scientist with the 2017 DM map project, has stated that, "The big question is whether Einstein's theory is perfect. It seems to pass every test but with some deviations here and there. Maybe the astrophysics of the galaxies just needs some tweaks. In the history of cosmology there are examples where problems went away, but also examples when the thinking shifted. It will be fascinating to see if the current 'tension' in Cosmology will lead to a new paradigm shift" [see Website Link 1].

If such paradigmatic shifts are to be realised, there is the need for pure scientific inquiry and the vital importance of encouraging creativity, imagination, possibly in the form of "out of the box" thinking. Such original thinking is important to scientific innovation and societal progression of our understanding of the universe and its dynamics.

The key question that is addressed by UE-GLEW theory is "How does the Universe work?" (I have referred specifically to the Universe that we occupy, and are transient guests within, through the use of a capital letter. Other universes within the Multiverse, have referred to with a lower case 'u').

Barriers to our thinking: concepts that inhibit our creative imagination and understanding of the universe, and other ways of thinking about the dynamics of the Multiverse

Our understanding of observable and unobservable phenomena is based upon our interpretations as hermeneutics or, more precisely, philosophical hermeneutics. That is, we create sense and meaning by inferring what leaps in the gaps between the observable and unobservable using a range of concepts. Therefore, we interpret what we can and cannot observe through our imagination, creativity, the introduction of new concepts, and intellectual insights (through prior knowledge and understanding of that knowledge). We are limited only by our ability to observe, interpret and imagine what is unobserved (or unobservable).

Einstein wrote of a realist approach to science, and our understanding of what we believe we know! Hawking (2008) stated that, "…if the universe is expanding, there may be physical reasons why there had to be a beginning" (p. 11). He states that it would be meaningless to suppose that it was created before the big bang (Hawking, 2008, p. 11). Indeed, there have been claims that "…events before the big bang can have no consequences, so they should not form part of a scientific model of the universe. That is, that we should, therefore, cut them out of the model and say that time had a beginning at the big bang" (Hawking, 2008, p. 26). However, such assertions are unhelpful, and, in themselves, barriers to developing ideas and new thinking that leads to further enhanced understanding and insights into the dynamics of the Multiverse and individual universes. Indeed, it has been stated that, "An important lesson of relativity is that there is less intrinsic in things than we once believed. Much of what we used to think was inherent in phenomena turns out to be merely a manifestation of how we choose to talk about them" (Mermin, 2005, p. 186).

(Hawking, 2008) argued that we can only look to find answers to the 'how' questions regarding the mechanics of the universe, as it is impossible to philosophically determine 'why'. However, Einstein was puzzled by the thought that the universe behaved without strict causality or certainty. He discarded Newton's concepts of absolute space and time but continued to embark on a quest for absolutes, certainties and invariants creating an orderly and dynamic mechanism underlying the laws of the universe (see Isaacson,, 2007, p. 3).

Einstein's theories and concepts were described, by Max Planck (1858 – 1947), as exceeding the audacity of the original thinking that previously been achieved in theoretical and speculative science. However, such audacious and original thinking led to a revolution in terms of how the mechanical systems of the universe and the Earth were perceived. Einstein was, like many scientists today, including me, "…a scientific realist …who believed that an underlying reality existed in nature that was independent of our ability to observe or measure it" (Isaacson, 2007, p. 169). Physical realities and events take place regardless of whether we can observe or measure them, or, indeed, that we are aware such events are taking place at a given time in a certain place. Reality lies beyond our perceptions, and, therefore, we must call upon imagination, interpretation and conjecture to fill in the gaps between our experience and reality.

Einstein asserted that gravity was responsible for the curvature of space through the interplay between matter, energy and momentum (p. 4). However, the principles of quantum mechanics need to be complete before it can be included within a unified field theory that includes QED, light and gravitational fields, and encompasses the curvature of space and the bending of light. The combination of gravity, force, mass and momentum in the one simple, elegant theory of UE-GLEW would enable a step towards making our thinking more complete, if not whole. That there is no such phenomenon as absolute time, as discussed across Einstein's published works, has the potential to be confusing as time is inevitably definable by observations and perceptions alone (p.82). In 1882, Planck had, however, referred to the need to assume that the universe consists of matter in a continuum. However, this requires having "…the brashness needed to scrub away the layers of conventional wisdom that were obscuring the cracks in the foundations of physics… [making] …conceptual leaps that eluded more traditional thinkers" (Isaacson, 2007, p. 93).

The principle of relativity was not a new discovery by Einstein: it had been a factor of mechanics written about by both Galileo and Newton. Einstein applied relativity to mechanical and electromagnetic phenomena (1905), and, subsequently, gravitational phenomena (1916). Indeed, it has been argued that, "…relativity is an …invariance principle (Mermin, 2005, p. 2). Furthermore, although an aside, Heisenberg's Uncertainty Principle (1927) may not directly apply to universal quantum mechanics through our lack of knowledge regarding the simultaneous position and velocity of a molecule does not change our understanding of black holes, the big bang theory and the movement of light within UE. Therefore, the uncertainty principle, whilst philosophically very clever, is a further potential barrier to developing our creative thinking and potential understanding of how the Universe works.

Einstein wrote and spoke of inertial mass and gravitational mass as equivalents within his equivalence principle, with the local effects of gravity and acceleration being the same. That is, "…the effects we ascribe to gravity and the effects we ascribe to acceleration are both produced by one and the same structure" (Janssen, 2002). It is interesting that Einstein refers to one of his famous 1905 papers as a heuristic: by this, he was referring to his presented ideas as a hypothesis that does not provide any form of proof but can only guide and direct the reader as to how a problem may be solved. Einstein later stated that, despite fifty years of pondering, he was no closer to determining what light quanta are (Isaacson, 2007, p. 101). Einstein's general theory of relativity incorporated accelerated momentum within the theory of gravity and gravitational fields. The key principles of relativity relate to objects and systems that are moving at a constant velocity relative to each other (Einstein, 1905b). However, for Einstein, there was the dilemma that objects do not always move at constant velocities in relation to each other (Einstein, 1916). There is always room for predictability and uncertainty due to changes and responses to these changes within the universe.

Combining fundamental physics and philosophy as a means of understanding through interpretation and perception of what has not and might not ever be seen! Indeed, Einstein's theories have been epistemologically probed, and continue to be debated in philosophical circles (see Isaacson, 2007, p. 315).

Einstein asserted that light waves involve momentum due to the disturbance of an unseen medium, as opposed to empty space that contains nothing. There has to be a medium that enables light waves to be transported akin to the way that sound waves are transmitted by air vibrations. That is, "...light must be interpreted as a vibratory process in an elastic, inert medium filling up universal space" (see Isaacson, 2007, p. 111). Einstein regularly stated that, "There is no real reason – other than either a metaphysical faith or a habit ingrained in the mind – to believe that nature must operate with absolute certainty ...some things simply happen by chance" (Isaacson, 2007, p. 334). However, Einstein was reluctant to give up complete causality. It was more about "...beautiful and subtle rules that determined *most* of what happened in the universe, whilst leaving a few things completely to chance" (Isaacson, 2007, p. 334).

Einstein began each of his 1905 papers with a discussion of oddities and incongruous ideas caused by jostling, incompatible theories. For example, light coming from any star seemed to arrive at the same speed, with the speed of a light beam being constant no matter the momentum of the light source. This thought troubled Einstein as such a postulate seemed to be incompatible with the concepts central to relativity. Einstein stated that, "Time cannot be absolutely defined...", and, therefore, there is no way to declare two spatially separated as truly simultaneous (see Isaacson, 2007, p. 123). Indeed, all moving objects have their own relative time. The same was asserted by Einstein for the removal of the long-held concept of absolute space and distance, as this was not always predictable. Space is not at absolute rest, which creates the potential for change. However, Einstein was adamant that the speed of light was absolute and invariant.

Maxwell's electromagnetism equations remained unchanged by Einstein but were extended to include waves, particles and emissions. Light quanta are a description of detectable light as it is emitted from the waves and medium that it travels in: to put this another way, its transition from undetectable light, travelling within UE, to detectable light that is emitted and absorbed.

Time was introduced as a fourth dimension by Hermann Minkowski (1864-1909): it is known as Minkowski space or Minkowski spacetime. It was originally introduced by

Minkowski as a means further understanding Maxwell's equations for electromagnetism, and then became associated with the postulates of special relativity (Einstein, 1905b). in 1907, Minkowski stated that Einstein's special theory of relativity could be understood geometrically as a theory of four-dimensional space–time (Minkowski, 1907). However, Henri Poincaré (1854-1912) had questioned the defence of the absolute nature of time.

Within this book, new thinking and concepts are suggested for discussion and debate by scientists and the general public alike. These are collated as UE-GLEW theory. This novel theory centres upon Universal Energy (UE) as the basis of all matter, energy, momenta and gravitational effects within our Universe. UE will never be directly observed at the quantum level but the impact of UE can be imagined and interpreted based upon all that we are able to observe. There are several approaches that we can take to thinking about the quantum components of the universe, if we are to understand their mechanical activities and consequences. As with different states of matter on Earth, dark matter exists in several different states, or forms, simultaneously. These different states are dark matter, gravity, light and dark energy. All four are simultaneously known as Universal Energy.

Four radical concepts that act as barriers to understanding the nature and functioning of all universes.

There are four key concepts that continue to act as barriers to understanding the nature and functioning of all universes. The first is the use of the word 'time'. The second is the long-held assertion that the speed of light is absolute at 300,000 km per second. The third concept that needs to be viewed differently is gravity: as an effect, rather than as a force. The final paradigmatic shift presents black holes as being of two different forms, acting as thermodynamic regulators, and, in the case of true black holes, as thermodynamic gateways to other universes.

If we are to understand how our Universe works, these are four changes to our thinking that we could put into place as means of removing conceptual barriers. We, therefore, need a scientific and societal shift in how we think about and seek to understand the invisible processes and mechanics that determine how the universe works. To make such a scientific and societal shift in our thinking will help us to develop a unified field theory. The fact that we have been unable to create a unified field theory to date is because we have been thinking about the wrong things or have been hanging on to concepts (and definitions) that prevent forward movement in our understanding.

Conceptual Revision 1 - Using the term *change* instead of 'time'

The Universe and Multiverse are not governed by the human concept of 'time' and are not confined by the principle of time. Time is a human construct to aid understanding and planning. Instead, the Universe and Multiverse are driven by change: changes to physical, chemical and natural processes that are altering and stabilising. Therefore, the first revolution within our thinking is to no longer use the term 'time' in relation to the quantum mechanics and dynamics of the universe: instead, we need to use the term '*change*'. The Multiverse functions and progresses according to *change*, rather than the constraints and perpetuation of time. Such change is manifested as the regulation, relocation and recycling of Universal Energy (UE) through GLEW streams, black holes and regulatory generation by stars. Change corresponds to the laws that are made unpredictable by physical objects across each universe. That is, the movement and position of physical objects is predictable as far as their compliance with the laws enable: therefore, not all laws are absolute in that such movement creates unpredictability.

Our ideas and understanding are informed by our chosen frames of reference. Within the theory of UE-GLEW, it is postulated that time should not be included as a frame of reference. Instead of time, it is argued that a more reliable frame of reference would be change and stability through Universal Energy. Time is the necessity for change and change is the reason for time. However, the concept of 'time', absolute or otherwise, does not apply to our Universe and other universes. Only humans need to be aware of time and the passing of time: time is the universal human obsession as it is the basis of the many forms of survival. The universes do not need time, for their whole focus is upon change.

Time, as we define and use the construct, is a human invention that brings some level of situational order and priority within an otherwise unordered existence. That is, time is only passing at the rate at which an individual is aware of passing the time; this perception includes the nature and progress of an activity that the individual is engaged in at present and is anticipating in terms of further development. In other words, time always proceeds at a

Figure 1   Key principles of UE-GLEW Theory (as of December 2021)

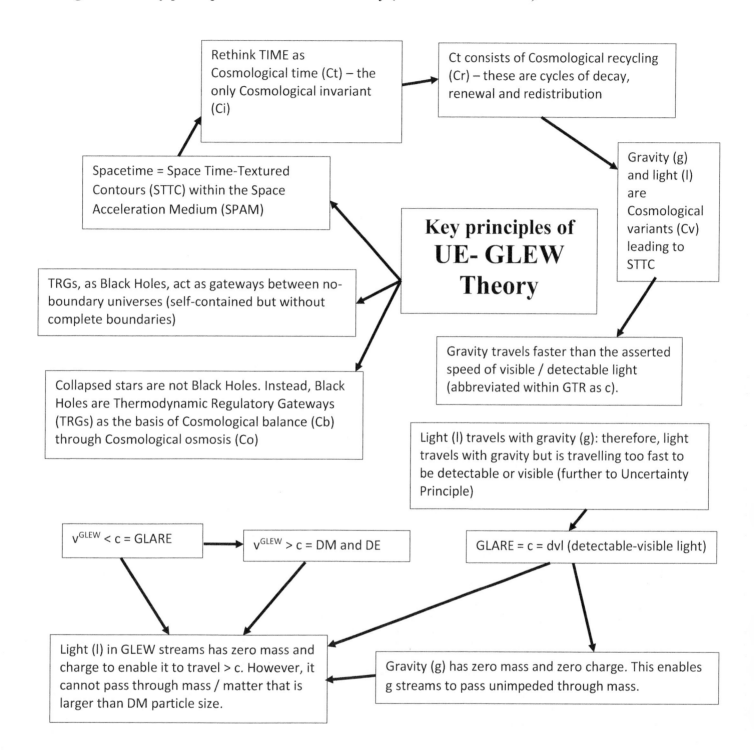

pace that is inverse to the pace at which we would like time to pass! Therefore, time is the universal connection between people, locally and globally, but not within the space fabric of the Universe and other universes. Time is human, created as a philosophical, psychological and mathematic construct, that acts as one of the key universal informants of human behaviour. There is, therefore, a time for change by altering time to *change*!

The term *event* plays a fundamental role in the relativistic description of the world but not, I contend, in the Universe, as "An event is something that happens at a definite place at a definite time" (Mermin, 2005, p. 45). However, this is an idealization because, as with Heisenberg's uncertainty principle, viewing something as an event depends upon our awareness of the event and how it is perceived in terms of spatial and temporal references (Ibid).

Whilst the term 'change' is used within the Theory of Universal Energy and GLEW (UE-GLEW), much of the change that occurs within the universe either cannot be observed or is so far away from us, as the observers, that we do not discern changes. Therefore, we cannot measure all changes within the universe, and, therefore, time does not exist as a measurable entity. Time is a human frame of reference, as a cause, in equal parts, for both understanding and confusion. However, they enable us to only understand our own existence, and confuse our perceptions and need to make clear decisions.

Time has been a barrier to the creativity of further thinking and the development of our understanding of how the universe works since Newton and Einstein. Even before that, Aristotle asserted that time is the measurement of change, with all things changing continually. Of course, change is invariably measured by time (Rovelli, 2019, p. 56). When Einstein introduced the 'fourth dimension' of 'time' as a basis for clarifying and justifying many of the concepts central to his theories of relativity (Einstein, 1905, 1916), he, unwittingly, formed an ongoing barrier to further thinking 'outside the box.' Of course, people like predictability and invariably thrive on stability. Through this, they develop a blinkered faith in perpetuation and certainty. They are perturbed by unpredictability and probability, which leads to the need to find reasons (causes) before deciding how to respond to human dilemmas and contexts. Invariably, these perceived causes and the resultant decisions are complicated by a wide range of known and unknown variables, one of which is usually time as an organisational structure and driver. However, time can be as much of a barrier to human thinking as it is an enabler or motivator. If we remove the concept of 'time' as we understand it, in human terms to define human activity and history, we can think about change in other ways in relation to the quantum mechanics, observable as larger scale movements of objects, within our Universe.

Universal Energy is composed of gravity, undetectable light, dark matter and dark energy. All four components of UE are all one and the same simultaneously. The frame of reference for the mechanics of the universe is *change* through non-judgemental adaptation, without interpretation or accompanying emotion. Redefining our notion of *time* as *cosmological change* will enable us to forget about chronological time as a human interpretive concept within cosmological dynamics and the behaviour of the universe. Otherwise, the confusion between human time and cosmological change will remain as a barrier to new thinking and the unification of the mechanics and processes of the universe.

Conceptual Revision 2 - The variable speed of undetectable and detectable light

The second revision is the postulation that light can and does travel faster than the know speed of light (abbreviated as c). The speed of light, represented by c, is, more accurately, referred to as the speed of light in a vacuum (Mermin, 2005, p. 25). However, space is not a vacuum as it contains UE, consisting of both DM and DE. In 1905, Poincaré had first proposed the idea of gravitational waves emanating from a body and propagating at the speed of light. Mass and energy are different manifestations of the same thing. Therefore, dark matter (dark mass) and dark energy are postulated as the same thing.

By rethinking and redefining our steadfast ideas regarding time and the speed of light, and, instead, using the term '*change*' and regarding light has being able to travel faster than 300,000 km per second, it is possible to move towards developing a unified field theory. Light that travels faster than the speed of detectable light, including visible light, shall be known as *undetectable light*, as is simultaneously part of the dark matter and dark energy that constitutes Universal Energy (UE). Undetectable light is abbreviated as ^ul and its speed is abbreviated as ^uc. If we accept that detectable, measurable light (^{d}l) travels at the asserted speed of light (known as c, and relabelled as ^{d}c), the next stage in our thinking should be to assert that UE travels faster than ^{d}l at ^{d}c. UE travels at ^{u}c, and consists simultaneously as gravity, dark matter, dark energy and undetectable light (^{u}l). In summary, light, as detectable light (^{d}l), is variant and inconsistent in its speed (^{d}c), and is not always predictable in terms of its transit through, for example, the atmosphere of planets and as measurably released by stars. Therefore, ^{d}l is variant and not always predictable from one moment to the next: ^{d}l is undergoing constant change in its velocity, frequency and intensity.

Conceptual Revision 3 – Gravity as an effect-response of UE rather than a force

Change enables stability and predictability as far as is possible when one takes the movement of physical objects into account. The momentum of physical objects creates instability and unpredictability, which UE responds to maintain stability and predictability. That is, to create and sustain order where there are constant opportunities for chaos to prevail. Gravity, as UE, travelling as DE and DM, creates gravitational 'pressure' to maintain stability within the universe, and a simple predictability to the movement of large objects moving through the spacefabric curvature (SFC). Of interest is the suggestion, by Hawking (2008), that there are forces that "…did not come from any particular source, but was built into the very fabric of space-time" (Hawking, 2008, p. 19).

The predictable momentum and velocity of large objects within the universe is due to UE pressure (UEP). Further to Newton's third law, the objects will create an opposite reaction to the UEP: however, although it is an opposite reaction, it is not equal. Therefore, Newton's third law is only partially applicable as, otherwise, all objects within the universe would become stationary. Indeed, Isaac Newton (1642 – 1727) posited that gravity travelled faster than detectable, measurable light, and that, therefore, there were phenomena that could travel faster than the speed of such light.

Through Universal Energy (UE), the universes are able to sustain the predictability of the movement of objects through the spacefabric, and ensure the perpetuation of change caused by such movement, in the form of spacefabric curvature (SFC). Order and stability enable predictability of momentum, increasing the order of the universe and reducing the probability of chaos. Such predictability, order and stability throughout the universes is due to changes, as adjustments and adaptations, caused by UE, travelling as GLEW streams, with regulation, recycling and relocation powered by black holes as thermodynamic regulators.

Conceptual Revision 4 - The form and functions of two types of black hole

Black holes have two different forms and functions. The first are collapsed stars, which create black voids as opposed to the 'black hole' that we might usually imagine within the spacefabric. The term 'black hole' was coined by John Wheeler (1911 – 2008) in 1969. Hawking (2008) describes black holes as "… a region of space-time, from which it is not possible to escape to reach a distant observer" (p. 41). Ambiguously, Hawking (2008) states that "Such objects are what we now call black holes, because that is what they are – black voids in space" (p. 35).

Black holes have a number of simultaneous functions. One function is to act as thermodynamic regulation gateways (TRGs) between universes. Their role, as TRGs, is to ensure the recycling, relocation and regulation of UE within a single universe and between universes. The second function of black holes is to be the source of the creation of new universes across an unknown, but infinitely large, number of universes within at least one Multiverse.

Further Discussion

All human experience unfolds relative to the time and space dimensions of the contexts we inhabit, have inhabited or about to inhabit. The *space* element of any context is the setting or location for an activity: this includes the presence or absence of other known human agents, the presence of unfamiliar human agents, the nature of the expected or perceived activity, and the resources available or known to be available in relation to the activity. The *time* element is essentially more complicated than *space* as much of it is unseen: for example, *time* could be a specific moment in the individual's life, including the development of cognitive processes and affective resilience; time allowed for the activity to be undertaken, and the possibility of revision

19

and / or repetition; the time of day (or similar timeframe, such as week, month or year) or the particular moment in time amongst other activities, and; the particular juncture of the development of ideas (whether someone is new to a particular concept or is well-versed in and understands the development and application of a concept).

We are located both physically and mentally in time and geography, with both having an impact upon the energization and nature of our perceptions and behaviours. The mind is physically in the present and experiences its environment in real time: the 'here and now'. Therefore, we have no choice other than to experience through geography (space, context and setting) and time (the moment in time that we are currently occupying and progressing through). However, mentally, the mind influences behaviours and responses based upon the past and the future. The individual mind takes the perceptions and motivations that are formed within the current moment (the 'present') and interprets them based upon past experience and future intentions.

Past experiences, of other individuals, contexts and successes, are formed through a combination of interpretations that are influenced by the combined impact of confidence, motivation, familiarity and expectations. How we come to regard our experiences is based upon a combination of the perceptions that we are either aware of or deliberately choose to filter through our interpretations and the contexts within which our experiences unfold. Perceptions are the outcomes of attention; what we choose to focus upon and what we choose to ignore, rather than the perceptions that we are unaware of. We may be unaware at the time that certain perception-based information was either not available to us or that we subconsciously chose to ignore such information. That is, perceptions are based upon attention: whether we choose to pay attention or we choose to actively ignore perceptions by directing our attention away from such perceptions.

Time is an essential factor at the heart of human experience, as outlined in the Philosophy of Confidence-Informed Social Motivation (PCISM: Wood, 2020, 2021). We are all travellers through time in that we have a finite amount of time in which we both exist and seek to live through our isolated and shared experiences. For the purposes of PCISM, 'time' has been viewed in a number of ways. Firstly, it regarded as something that is forward moving and unstoppable, in that whilst we are aware of the past and are preparing for the future, we constantly exist in the present. Secondly, time is regarded as something that is relative to the activity and the 'space' (including the presence of other human individuals). This is likely to regarded differently by individuals in the same context and space, according to the perceptions, expectations and confidence levels of each individual.

CHAPTER 2

The integration of the four cosmological revisions as the basis for UE-GLEW Theory

Rethinking our thinking: towards a Theory of Everything becomes Everything Else!

As previously mentioned, the key question that is addressed by UE-GLEW theory is "How does the Universe work?" UE-GLEW is the abbreviation for Universal Energy: Gravity-Light Energised Waves.

To provide some of the answers, the remainder of this chapter presents and discusses a *Theory of Everything about Everything Else,* in the form of UE-GLEW as a proposed unified field theory. There are four key assertions at the heart of the UE-GLEW theory. Acceptance of these assertions will require the majority to rethink their understanding of the dynamics of the universe. All four of these concepts revolve around the assertion that **the Universe is a dynamic, momentum-driven mechanical system, thermodynamically regulated and linked to other universes by black holes.**

   1. **Universal Energy (UE) as the basis of all matter, energy, momenta and gravitational effects within our Universe.**

The first radical concept that the UE-GLEW theory centres upon is that **Universal Energy (UE) as the basis of all matter, energy, momenta and gravitational effects within our Universe**. UE will never be directly observed as quantum particles but the impact of UE can be imagined and interpreted based upon all that we are able to observe. There are several approaches that we can take to thinking about the quantum components of the universe, if we are to understand their mechanical activities and consequences. When thinking about how the Universe works, concepts and theories tend to be centred upon gravity (g) and gravitational fields (GF). However, within UE-GLEW, **gravity (g), light (l), Dark Matter (DM) and Dark Energy (DM) are combined as Universal Energy (UE)**, expressed as:

UE = g = ^{u}l = DM = DE, where ^{u}l = undetectable light

UE travels as Gravity-Light Energized Waves (GLEW) through and as a result of DM, DE, g and l. Given that light is neither emitted nor absorbed whilst it is travelling as part of UE, it is known as GLUE (Gravity-Light Universal Energy). Therefore, GLEW is detectable and measurable whilst GLUE cannot, to date, be detected or measured.

UE is passing through and around us in the same way that it passes through and around much larger objects, including planets, stars and black holes. However, as human individuals, we have too small a surface area to volume (SAV) ratio to directly detect and feel the influence of gravity, as long as we remain stable and balanced!

As with different states of matter on Earth, dark matter exists in several different states, or forms, simultaneously. These different states are dark matter, gravity, light and dark energy. All four are simultaneously known as Universal Energy. Therefore, the UE-GLEW theory draws all four postulated conceptual revisions together, and are summarised in a simple equation:

$$\textbf{UE + GMR = ERIN + SPIN}$$

Where:

a. **ERIN** is an abbreviation of the phenomenon of Energy-Regulated Inertia.

b. **GMR** is an abbreviation of the phenomenon of Gravitational-Mass Resistance. Gravity is argued to be an effect in response to imbalance, instability and disorder within the momentum of objects.

c. SPIN is an abbreviation of the phenomenon of SPace INertia.

d. Black holes, in the two different forms of BH-TRGs and BV-SPIN-ERs, are the causal structures of **ERIN.**

e. **SFC** is the spacefabric curvature. In other cosmological theories, especially relativity, this is known as the spacetime curvature (stc). However, as the concept of time has been substituted with *change*, the label of 'the spacetime curvature' no longer applies.

This equation will now be unravelled and discussed. However, this requires changes to some specific long-held ideas about how specific aspects of the universe are defined and viewed. The most important of these are that:

1. There is no relative 'time' within the processes and mechanics of the universe. Instead, *change* is the cosmological constant and the fourth dimension.
2. Undetectable light (^{u}l) travels faster than c as the asserted speed of light. Detectable light (^{d}l) travels at the asserted speed of c or less than c, depending upon the medium it is passing through.
3. Gravity (g) is not a force: instead, it is an effect and a response that is detectable due to imbalance, instability and disorder in the momentum of objects: an effect-response.
4. Black holes exist in two different forms: BH-TRGs and BV-SPIN-ERs. These are the sources of **ERIN.**
5. **SFC** is the spacefabric curvature. In other cosmological theories, especially relativity, this is known as the spacetime curvature (stc). However, as the concept of time has been substituted with *change*, the label of 'the spacetime curvature' no longer applies.

In addition to **UE + GMR = ERIN + SPIN**, the above reciprocal and recyclical processes are summarised in the equation:

UE = GLEW = SFC + BHTR

$$\longleftarrow \hspace{3cm} \longrightarrow$$

That is, **GLEW = UE [DE + DM] as g and ^{u}l**, otherwise expressed as:

UE [DE + DM = g + ^{u}l]

Synonymously:

UE = DM + DE $\longrightarrow$ g + ^{u}l

24

ERIN + GMR = SPIN + SFC

The four simultaneous components of UE are the means by which stability, order, consistency and perpetuation are regulated features of the universe. SPace INertia is abbreviated as *SPIN*. SPIN is defined as the stability that is in place due to inertial object momentum, due to object rotation. The above concepts are summarised in Figure 1 (in Chapter 1) and Figure 2 below.

Figure 2

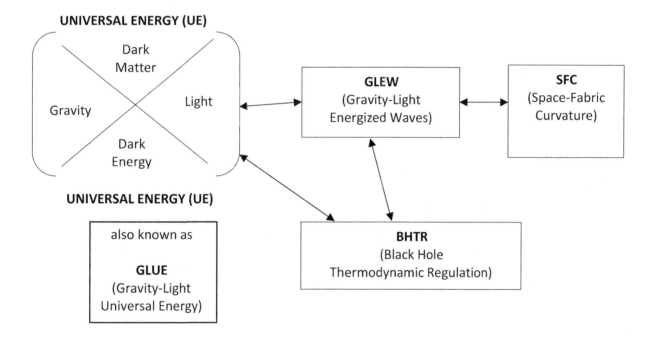

UE ensures that stability is maintained, and stability is restored when instability has caused unpredicted movements and changes to the path of objects within the universe. UE is predictable and consistent as steady streams of GLEW, carrying UE in its different and simultaneous forms of gravity, undetectable light (^{u}l), dark matter and dark energy. That is:

UE = g = ^{u}l = DM = DE = GLEW

UE travels as GLEW streams, directed by changes in the SFC, and, simultaneously, as a means of sustaining of SFC. Similar to, gravity influences SFC whilst the SFC determines how gravity acts in relation to the SF and objects within it.

^{u}l is the undetectable speed of light, which is undetectable in that it cannot be seen or measured as it is travelling faster than c (the asserted speed of light at 300,000 km per second). Undetectable light (^{u}l) is non-visible, is not emitted, and is not absorbed in the same way as detectable, visible light (^{d}l). This is because ^{u}l is a simultaneous constituent of UE: as has been postulated, UE travels faster than the speed of ^{d}l: that is, faster than c as the asserted speed of visible light. Undetectable light (^{u}l), travelling faster than c, becomes detectable light (^{d}l) when it is decelerates to c or less than c.

Gravity and ^{u}l travel as constituent components of UE as they are both dark energy (DE) and dark matter (DM) simultaneously. They are all the same variables simultaneously and individually. Therefore, UE > c. UE does not 'perceive' its own mass and acceleration. Therefore, it is not inhibited by the spacefabric curvature (SFC) as it is the SFC. That is, **UE = SFC.**

When we are stable (or is as close to stable as possible), we do not feel the influence of gravity. It is only when we become partially or completely unstable that we feel gravity as the stabilising force. Inertial frame of reference applies equally to something that is uniformly moving as much as it does to being stationary (Mermin, 2005, p. 4). The dynamics of the universe may be regarded as inertial, uniformly moving frames of reference, with the spacefabric being an equally inertial frame of reference.

UE is predictable and consistent in its power to create and sustain stability, unless something else causes an object to behave differently, such as becoming unstable or changing its path of momentum. This raises the potentially unanswerable question of whether gravity and UE are trying to move objects in predictable, orderly and sustainable paths of momentum, or they are endeavouring to create balance, inertia and the immovability of objects.

2. Using the term *change* instead of 'time'

The first conceptual revision, and shift, involves adopting the term *change* instead of 'time'. Time has been discussed in Chapter 1, in terms of its usefulness as a human philosophical and organizational construct. Time is both relative and subjective in terms of the interpretation of moments, instants and instances (Wood, 2020, 2021). To experience a context fully relies upon the mind being simultaneously guided within the *present* moment by the experiential influences of the past as well as the hopes, desires, goals and dreams that guide individuals towards the future. Therefore, it may be that the mind is very rarely, if ever, functioning entirely in the current moment of time. Instead, the mind will be functioning in the past, present and future simultaneously as the basis of all conscious interpretations and subconscious perceptions (Wood, 2020, 2021).

By replacing 'time' with *change*, and introducing the concept of Universal Energy (as gravity, light, dark matter and dark energy), and the functions of black holes, we can create a means by which it is possible to bring the quantum mechanics of the universe together, as defined and discussed by numerous theorists, to create a unified field theory. Or, as Stephen Hawking would have labelled it: a Theory of Everything. I have, rather wryly, opted for two terms: the Theory of Everything becomes Everything Else, and the Theory of Something and Nothing! That is, the predictability and stability of the momentum of physical objects across the universe is due to numerous 'somethings' which, due to being undetectable and immeasurable, appear to be 'nothing'. It is only through the observable and measurable movement of such objects, for example, that we know nothing is resulting in something! There are many elements vital to our sustained existence that cannot be seen but we still regard them as 'something' as we know that they are present. This includes air and gravity. We know that they are present, even though we very rarely consider their existence daily, for their absence would be regarded, at the human level, as catastrophic!

However, 'time' limits our thinking when it comes to the dynamics of the Universe. Therefore, *change* is asserted as the fourth dimension, rather than time. Indeed, change is the cosmological constant. Change is cosmologically constant as it is seeking to sustain inertia and momentum, through UE, and is the means by which gravity has change-adaptive effects that are able to restore stability and inertia. Remove the concept of time as use it mathematically, and everything within the universe potentially becomes non-relativistic according to the theory of general relativity (GR). However, this is not strictly true in that change, inertia and momentum are relative to other objects, the SFC, ERIN and SPIN.

It had been noted by Hawking (2008), that, "We don't yet have a complete and consistent theory that combines quantum mechanics and gravity" (p. 86). Richard Feynman, through quantum electrodynamics, proposed that quantum theory should be approached in terms of a sum over histories: however, this still incorporates 'time' as opposed to change as a frame of reference. In addition, "…time is imaginary and is indistinguishable from directions in space" (Hawking, 2008, p. 87). Indeed, Feynman approached time as 'imaginary', in that time was indistinguishable from space. That is, "…the so-called imaginary time is really the fundamental time, and that what we call real time is something we create just in our minds" (Hawking, 2008, p. 91).

Acceptance of this principle alone would reframe our thinking about time in relation to the dynamics of the Universe. In real time, the universe is regarded as having a beginning in the form of a singularity which, therefore, creates a boundary. However, in imaginary time, which is asserted as change as the new frame of reference, there are no singularities or boundaries (Hawking, 2008, p. 91). Regardless of whether one is drawn towards 'real time' or 'imaginary time', the use of the word 'time' still causes human confusion! Let us instead approach what happens within the mechanics of the universe as *change*.

The general theory of relativity (Einstein, 1916) is about time as a concept, defined by a series of events (Mermin, 2005, pp. ix and 171). An event is defined as something that happens in a definite place at a definite time (Mermin, 2005, p. 45). Einstein gained inspiration and developed his ideas through observing and reflecting upon human and natural phenomena around him. Einstein (1916) wrote that "…a time-value is associated with every event which is essentially capable of observation" (p. 24). Einstein (1916) recognised that space is a three-dimensional continuum, with the inclusion of time creating a fourth dimension (p. 55). Einstein stated that

"…before the advent of the theory of relativity, time played a different and more independent role, as compared with the space co-ordinates …treating time as an independent continuum" (Einstein, 1916, p. 56). With the theory of relativity, "…time is robbed of its independence" (p. 56).

The general theory of relativity is partially about observation and perception, which is, in turn interpretive (hermeneutic) based upon what we are seeing as the focus of our attention (Hanson, 1958). In other words, Hanson (1958) argues that what we see and perceive is not what our senses receive, but is instead filtered sensory information, where the filter is our existing preconceptions. This includes embracing sources of angst within scientific thinking, including uncertainty, incompleteness, and the undecidability and unprovability of some phenomena. This concept later came to be known as a 'thematic framework.' (Hanson, 1969, 1971). Therefore, we interpret our perceptions based upon our limited focus of attention and what we believe we know and understand. However, time and attention, as well as interpretations, are subjectively referential, depending upon our accepted frames of reference. By contrast, change and adaptation within the universe are non-judgement, non-emotive and absolute: it adapts only through cause and correction.

The *change* that is central to the regulatory dynamics of the Universe does not involve observation, perception, interpretation and decision-making: only the responsive mechanics of cause, effect and correction as a means of ensuring that the universe remains in as ordered a state as possible. Therefore, time has no place in the governing dynamics of the Universe. The only thing that matters is object momentum and the space within which the momentum takes place. When objects move in ways that unpredictable to us, as in they either unexplainable or did not happen in the way we had predicted, we try to explain and understand the 'how' and the 'why'. By contrast, the dynamic mechanisms of the universe adapt and change according to something that has changed to undermine its stability!

3. The variable speed of undetectable and detectable light

As stated, ^{u}l is the undetectable speed of light, which is undetectable in that it cannot be seen or measured as it is travelling faster than c (the asserted speed of light at 300,000 km per second). Undetectable light (^{u}l) is non-visible, is not emitted, and is not absorbed in the same way as detectable, visible light (^{d}l). This is because ^{u}l is a simultaneous constituent of UE: as has been postulated, UE travels faster than the speed of ^{d}l: that is, faster than c as the asserted speed of visible light. "… the velocity of propagation of light cannot depend on the velocity of motion of the body emitting the light" (Einstein, 196, p. 17) and "The velocity of propagation of a ray of light …comes out smaller than c" (Einstein, 196, p. 18).

UE is 'transparent', in that it cannot be seen as it is travelling at momenta faster than c (the speed of ^{d}l). Therefore, ul becomes ^{d}l when it decelerates to momenta of c or less. When ul becomes ^{d}l, due to the deceleration of UE to c, heat and light are emitted as an object has been encountered or the SFC (spacefabric curvature) leads to rapid deceleration. In summary, ^{u}l $> c$ whilst ^{d}l $= \leq c$. This explains why ^{u}l is undetectable and invisible, as it simultaneously exists as UE, in the form of DM and DE. At speeds greater than c, ^{u}l is not subject to the photoelectric effect: it is only subject to the photoelectric effect, including emission and absorption, when UE $\leq c$. This also explains why we cannot detect the momentum of light waves travelling through the darkness of the spacefabric. The spacefabric (SF) is composed entirely of UE (as g, ^{u}l, DM and DE simultaneously).

This leads to the rewriting of Einstein's formula (E = mc2) as $E = {}^dm \times {}^dc^2$.

Within the postulates of UE-GLEW theory, the formula central to the general theory of relativity may also be written as $UE = {}^um \times {}^uc^2$. This, in turn, can also be expressed in two ways that are equal:

$DE = DM \; {}^uc^2$

$E = M \; {}^dc^2$

In other words, $E = mc^2$ becomes ${}^dE = ({}^dm) \times ({}^dc)^2$ where ${}^dc =$ the detectable speed of light.

Light travels significantly slower in transparent media such as water and glass, and is slightly slower within air and atmosphere (Mermin, 2005, p. 23). Therefore, the constancy of the speed of light is counterintuitive as it varies with the medium it is travelling through (Mermin, 2005, p. 25).

4. Gravity as an effect-response of UE rather than a force

Universal Energy, change (rather than time) and undetectable light, travelling faster than the mathematical speed of light, are three radical posits that could help free our thinking in terms of how the Universe functions and regulates itself. However, the most radical conceptual change, potentially is that gravity has one function: as an effect that leads to outcomes that result in stabilisation, balance and predictability, as opposed to instability, imbalance and unpredictability.

Gravity is only perceived or detected when there an object becomes unstable or unbalanced. For the most part, UE passes through and around ordinary matter (OM). This transit of UE results in stability, balance and inertia of momentum as norms, within the human-known laws of physics. It is only when the inertial frame of reference is altered that stability and unbalance becomes obvious, with gravity acting as the effect, rather than a force, as momentum is stabilised (even though the trajectory and velocity of an object may now be different). Change involves a variance in the stability and balance of momentum of an object, something that is not directly influenced by or caused by UE.

Such changes and their consequential effects are part of the ongoing evolution of the universe. The universe is dynamic, and momentum-driven in that it responds to cause and effect. This leads to correction as far as dynamism and momentum allow. Rather than looking at predictable and unpredictable phenomena as 'cause, effect and correction' within the universe, we seek the 'how' and the 'why'. These are only potentially monumental to us in terms of the physical and mental impact that such phenomena may or will have upon our survival! By contrast, the universe does not seek to understand 'why' something has changed or is happening, but instead just solves what has occurred by responding within the constraints of the governing dynamics and mechanical processes that influence momentum.

One of the central ideas of the theory of general relativity (Einstein, 1916) was that gravity arises from the curvature of spacetime and that, simultaneously, the spacetime curvature is due to gravity, defining how it should move and curve. Spacetime curvature is a means of more effectively ensuring the conservation of energy and momentum through arbitrarily moving reference systems. That is, momentum is arbitrary, accelerated, rotational and inertial, at a consistent and orderly velocity. Such inertia, in the form of rotational and gravitational forces, is only relative to the momentum of other objects. It, therefore, becomes a gravitational interaction between masses, rather than necessarily being an effect of space.

Gravity has to begin somewhere and be perpetuated somehow. Space, as the spacefabric, is not independent of gravity, as gravity influences the curvature of space. This curvature determines further gravitational action in an ongoing, responsive and dynamic set of processes designed to ensure stability, balance and order within localised areas of the Universe.

Mass is invariably incorrectly defined. Mass has been defined as both the quantity of matter and weight. It may have been defined as weight, but this depends upon where an object is as the measurement of weight is dependent upon the force that gravity exerts on the object (Mermin, 2005, p. 144). The term 'mass' should be defined and regarded as "…a measure of how hard it [an object] resists attempts to change its velocity" (Mermin, 2005, p. 145). Alternatively, and equally applicable, mass "…is a measure of how little the velocity of the particle changes in such collisions…": that is, although informal, the bigger the mass of an object, the less its velocity changes (Mermin, 2005, p. 145).

Mermin (2005) asks why "…gravity should act more strenuously to change the velocity of more massive objects in just such a way as to lead to exactly the same changes in velocity, whatever the mass?" (Mermin, 2005, p. 171). The influence of gravity appears to be directed towards the centre of an object, such as the Earth. It would be wrong to claim the gravitational field is uniform: however, there are nonuniform gravitational fields (Mermin, 2005, p. 172). An interesting comment, worthy of further discussion, is "…a gravitational field is determined by the distribution of matter that gives rise to it" (Mermin, 2005, p. 172). In addition, "…all effects of this gravitational field on all physical phenomena are indistinguishable from how the phenomena would play out in the absence of any gravity" (Mermin, 2005, p. 173).

Galileo and Einstein both recognised the principle of equivalence, in that the effect of gravity on the motion of an object is independent of its mass. However, this equivalence principle only applies in a vacuum. A vacuum is artificially created, whereas the universe is not a vacuum. Therefore, it is better to define the mass of an object as a measure of how hard it is to change the velocity of the object, whether through gravity or collisions (Mermin, 2005, p. 171). The gravitational field is nonuniform and complex (Mermin, 2005, p. 172). This includes nonuniform gravitational fields such as those prevailing over the Earth. That is, "…a gravitational field is determined by the distribution of matter that gives rise to it" (Mermin, 2005, p. 172).

All objects and structures, including living beings, stars, planets and black holes (in both forms: see the next section within this chapter), are subjected to effects of gravity and the gravitational field due to the instability of moving objects. The larger, in terms of mass, and the faster an object is moving, particularly if it is accelerating, the more instable it will be: therefore, the influence of gravity as a stabilising force will be more easily felt as *gravitational strength*. Gravity has been asserted as a constant force: indeed, it is, potentially, the only constant force across the Universe and all universes within the Multiverse. It is only *detected, manifested* and its influence '*felt*' by moving objects that become unstable: at this point, gravity is detected and manifested as it creates stability, even if the consequences are regarded by humans as disastrous or damaging.

We are constantly resisting the influence of gravity, even though we do not perceive it (through adaptive thought and interpretation), we are subconsciously adjusting to ensure that we remain stable, and, therefore, are not unduly impacted by gravity's influence to move unstable objects back to stability. The larger an object is, allied with its mass, combined with its acceleration, the greater will be its instability and potential for deviation from a stable course of travel.

The inertial frame of reference refers to an object that is uniformly moving (Mermin, 2005, p. 4). That is, "…if the same object moves uniformly, it will have the same properties in a frame of reference that moves uniformly with it" (Mermin, 2005, p. 5). However, the concept of absolute rest has no meaning or applicability for objects in transit through the universe, electromagnetic phenomena or Newtonian mechanics (Mermin, 2005, p. 27). Interestingly, UE is the exception to Newton's third law of motion, in that gravity (simultaneously as UE and DE): that is, there is not an equal reaction between the momentum created by the gravitational field and the opposite reaction of a mass-rich object resisting a change in velocity (Mermin, 2005). If the action-reaction were equal, there is need for the object to spin as a means of offsetting the potential for non-movement (absolute rest). It is of course recognized that non-movement is not possible as there is no actuality of absolute rest (Einstein, 1916; Mermin, 2005).

The momentum and acceleration of an object in space is based upon UE moving through and around the object. Here, Newton's Third Law partially applies as equal and opposite forces are exerted by the object and by gravity / UE in the form of *gravitational (UE) pressure*. However, the law only partially applies in that gravitational (UE) pressure has to be greater than the opposite force exerted by the object moving through space. If the reactions were the same, the object would not move and would be without momentum through the universe. Therefore, UE is the exception to Newton's third law of motion, in that gravity (simultaneously as UE and DE): that is, there is not an equal reaction between the momentum created by the gravitational field and the opposite reaction of a mass-rich object resisting a change in velocity (Mermin, 2005). If the action-reaction were equal, there is need for the object to spin as a means of offsetting the potential for non-movement. It is of course recognized that non-movement is not possible as there is no actuality of absolute rest (Einstein, 1916; Mermin, 2005). The rotation of an object within the spacefabric curvature (SFC) is, therefore, the means of ensuring that over-acceleration is prevented and, therefore, stability is more assured, whilst instability is less likely to occur.

GMR is an abbreviation of *Gravitational-Mass Resistance*. In this case, mass is defined by Mermin (2005), asserting that mass is a measure of an object's resistance to changes in its momentum. That is, that all objects seek an inertia of momentum that is resistant to change. Therefore, following such reasoning, gravity is a form of mass resistance (or GMR).

From this, given that I have already considered the use of '*change*' as opposed to time, this would be an appropriate juncture to introduce a new, and, potentially, more radical idea that might enable greater creativity when thinking about the quantum mechanics and processes of the universe. That is, that gravity may not be a force. Instead, it may be an effect caused by UE. Therefore, UE is the cause whilst gravity is an effect, manifested as a detectable outcome. Ultimately, we are rethinking how we define both gravity and light, as being either detectable or undetectable. Within UE-GLEW theory, gravity has been considered from the perspective of being an effect rather than as a force.

5. Black holes: The form and functions of two types of 'black hole'

UE travels as GLEW streams, with the recycling (regeneration), regulation and relocation (momentum) of UE being due to, primarily, black holes. Black holes have a thermostatic function as thermoregulation gateways: their function is to ensure that the gravitational pressure (GP) (also known as universal energy pressure: UEP) of the universe is thermodynamically regulated through dark matter and dark energy. To enable black holes to sustain GP / UEP, a single black hole would not be sufficient as a TRG. Therefore, black holes do not function and generate UE in isolation: instead, a huge network of black holes would be needed, similar to the holes in a string bag!

Similar to light existing and functioning simultaneously in two ways, as waves and particles (or detectable quanta), black holes take two different forms but with similar functions. The first form of black hole, as coined by John Wheeler, originates with the collapsed star. However, as Hawking (2008) has stated, these are not really holes as such: instead, a collapsed star creates a 'black void'. Each black void functions by regulating the distribution of UE through relocation. For the purpose of this theory, I have labelled these black voids as BV-SPIN-ERs (Black Void SPace INertia – Energy Regulators).

The second type of black hole are, indeed, holes, as discussed in Wood (2020). These are TRGs (Thermodynamic Regulating Gateways) between universes. Such TRGs function, between universes, as means of UE regeneration and recycling, regulation and relocation.

Therefore, if we are to fully understand black holes and their functions, we need to think of them as TRGs and SPIN-ERs.

Given that labels, once applied in physics, are invariably sustained, it may be better to write about these as BH-TRGs and BV-SPIN-ERs. BH-TRGs have the potential to be the reason for the creation and expansion of universes. A key puzzle that arises regards how such BH-TRGs are created, in order to create holes in the spacefabric.

The direction of rotation of both BH-TRGs and BV-SPIN-ERs determines the direction of the movement of UE, whether left-handed (counter-clockwise) or right-handed (clockwise). All rotating objects within the universe rotate with a given angular momentum. This rotation is a means of ensuring gravitational-mass resistance (GMR). Planets, asteroids, stars, solar systems and galaxies are all rotating due to *GMR-SPIN*. Indeed, it appears likely that the universe is rotating, as UE, in either one direction or may, due to separability and localization, be rotating in all directions according to the rotational direction of nearby BH-TRGs. With both BH-TRGs and BV-SPIN-ERs, it will not be possible to 'see' the direction of rotation as the UE consists of ^{u}l (undetectable light) travelling at velocities faster than c.

Therefore, given that it is known that all stars rotate, this suggests that 'black holes', as they are conceptually envisaged, also rotate if they are formed from collapsed stars: that is, it is expected that all BV-SPIN-ERs are rotating as a means of sustaining GMR-SPIN.

It appears that objects, consisting of normal matter, rotate but only rotate due to the large-scale and localized rotation of the universe. CMBR does not provide any direct evidence of universe rotation: however, it may be that CMBR, emitted as ^{d}l as an output of UE, is not measurable or detectable at the current time as it still exists as ^{u}l.

The regulation, recycling and relocation of UE, through the combined functioning of BH-TRG and BV-SPIN-ER black holes and stars, is directly causal upon the curvature of space fabric and texture. If UE travels through a continuous medium, it will travel as a wave that has an impact upon the curvature and texture of the space fabric that also consists of UE. The second law of thermodynamics has been detected as variable. In addition, it may be Newton's third law of motion is made invariant by the action of rotation. The action of rotation results in SFC and the torsion of both types of black hole.

The aforementioned dark matter map has created a further mystery (see Chapter 1). That is, what are the dark voids around the light halos caused by galaxies? Could these be BH-TRGs, and, therefore, actual black holes as opposed to voids? It may be that the voids are BH-TRGs as such but are indicators of the location of BH-TRGs far beyond what can actually be seen. The black voids are unlikely to be singular collapsed stars (BV-SPIN-ERs) as the dark areas are too large.

There is the potential that, from the Earth, an individual will be looking outwards towards the black voids. Thus, we are looking *towards* to the edges of the universe, although we cannot detect the actual boundaries as these remain beyond our capabilities to know when we have seen them. This is very similar to the assertions of Dr. Nikodem Poplawski, of the University of New Haven. There is, therefore, the possibility that, further to Poplawski's ideas, we are actually inside the product of one or more BH-TRGs where black holes have led to the creation and evolution of our own universe. The black voids that can be seen on the dark matter map may, therefore, be more than one BH-TRG concentrated in the same location. By looking outwards towards the boundaries of the universe, it may be that singular BH-TRGs are non-detectable due to their concentration.

CHAPTER 3

GLEW: the theory underpinning UE-GLEW Theory

In 2019 and 2020, the antecedents of this new UE-GLEW Theory were published, as books and a scientific paper (Wood, 2019, 2020). The two editions of the book, 'The new Big Bang Theory, Black Holes and the Multiverse explained' presented Gravity-Light Energized Waves as the GLEW holding the Multiverse together. It was stated that, if such a theory was to be held as plausible, we would need to rethink the composition and function of black holes, dark energy and dark matter.

The central ideas of GLEW Theory (Wood, 2019, 2020) are outlined and discussed below, as a means of clarifying some of the new posits within UE-GLEW Theory. After centuries of discussion and ongoing debate, it is suggested that if we are to get even remotely close to formulating a unified theory of what gravity 'is' and, as a result, how the structure of the Universe, and the wider Multiverse, is influenced by gravity, mass, heat and light, we need to abandon some of the thinking that we cling to so dearly as given or immovable 'facts'. To paraphrase Einstein, we cannot hope to solve the puzzles that intrigue and elude us with the same thinking we used when we created them. If we are constantly unable to find answers to our questions, it suggests that we are never going to find the answers using the theories and models that we currently rely up. In short, we may need to rethink the ideas and measurements that we hold as constants and relook them by reconsidering what may not always be 'solvable' or proved by mathematics.

The key proposal within GLEW theory, as it stood in June 2020, was that the quantum particles of gravity (gravitons) move faster than the accepted speed of light photons. Gravity is asserted as the smallest of all particles (known, undiscovered and never to be discovered) within all quantum and cosmological theory. Gravity particles are constantly interacting with other fundamental particles in order to maintain balance and order within the Multiverse. When gravity travels at the speed of light, it is both carrying photons and taking them to a speed where they are neither visible nor detectable: gravitons and photons travel together as Gravity-Light Energized Waves (abbreviated to GLEW, pronounced glue). That is, non-detectable photons travel faster than the asserted mathematical speed of light, expressed as c. When GLEW streams decelerate to the point where photons are travelling at a velocity that light becomes detectable, this is the point at which the Gravity-Light Acceleration-Related Energy (GLARE) threshold velocity is achieved.

The formulae for calculating the velocity of GLEW streams as they move through the universe, and between universes through black holes, are proposed as:

$$E^{GLEW} = \frac{v^{GLEW}v^m p V^0}{G}$$

$$v^{GLEW} = \frac{E^{GLEW}v^m p V^0}{G}$$

As *every GLEW stream travels faster than the speed of detectable light*, gravity was centralised as both the restorer of and maintainer of cosmological balance and order. Such balance is maintained by the movement of dark energy moving between black holes: in this way, black holes act as Thermodynamic Regulation Gateways (TRGs). Black holes, acting as TRGs, provide the mechanism by which gravity maintains multiverse-wide balance, with changes in the velocity of gravity being due to an inescapable imperative to maintain balance across the Multiverse. By sustaining balance, and, thus, both preventing and rectifying imbalances, gravity, therefore, must travel at different speeds within our Universe and the wider Multiverse.

In consequence. the ideas central to GLEW theory ultimately lead to the assertion that the Standard Model of physics needs to evolve and think again if we are to move forward with developing our understanding of cosmological and quantum concepts that have, to date, eluded us. Therefore, modifications to the way we think about the gravitons and photons, together with the resultant movement and impact of light and gravity, are needed if we are to move forward with our understanding of the gravitational and allied thermodynamics processes that ensure the stability of the Multiverse.

This novel theory of *Gravity-Light Energized Waves* (abbreviated to GLEW: pronounced as glue) outlined within this paper discusses a different view of the mechanics of gravity, including the curvature of light, black holes, dark matter and gravitational waves, without resorting to mathematical symbols. This theory has evolved the ideas of Newton, Einstein and Hawking. The ideas of both illuminated the 'what' of gravity when conducting their 'thought experiments'. However, the debate continues about the mechanics of 'how' gravity acts, together with the underlying 'why' relating to gravity behaves and influences as it does.

This new theory places gravity firmly at the centre of the key phenomena of our universe and the multiverse through dark matter, dark energy and black holes. There is a predictability to the way that objects and light moves within the Universe, a predictability that has enabled successes within space travel and the determination of the movement of objects such as comets and asteroids. Einstein famously stated that God does not play dice with the Universe. Instead, as asserted by Newton, there has to be a 'clockwork' logical predictability to the momentum of objects within the universe. However, this does not mean that the processes and products of the many universes that combine to form the Multiverse are stable and in a constant state of order. Instead, the gravitational constant, together with thermodynamic processes and the conservation of angular momentum, ensure that there is a constant correction from imbalance and disorder to balance and order. This process is perpetual and, from a measurement and detection perspective simultaneous. Gravitational and thermodynamic processes need to be considered across the Multiverse and between universes if they are to be fully understood in terms of their role in maintaining order and balance of heat, light and momentum, rather than at the 'localised' level of what is happening within the Earth's atmosphere. Gravity on Earth is nothing more than the outcome of what is happening far beyond our atmosphere!

Ultimately, the outlined theory of GLEW and GLARE builds upon prior 'thought experiments' by combining quantum mechanics, general relativity and (super) gravity, alongside, contained thermodynamic systems, (super) symmetry and breaking (super) symmetry through a number of simple premises about gravity and light. It does so by seeking to explain an inseparable relationship between gravity and light, the invisibility of light as dark energy and within dark matter, and both the variability and predictability of gravity within the boundaries of known galaxies, universes and black holes.

Light and the theories of relativity through the lens of GLEW Theory

One of the central principles of the theories of relativity was that the speed of light is constant. However, GLEW theory proposes that such a definition needs to be revisited and rethought; that is, that the perceived speed of light is only the speed at which light can be detected and is made visible. That is, that the speed that which light travels is *not* constant. When photons are travelling as part of a GLEW stream, they will be travelling at a speed faster than the asserted speed of *detectable* light. Whilst travelling at a velocity greater than the speed at which light can be seen and is detectable, bonds are sustained between the constituent GLEW particles: therefore, the bonds will be too strong for the photons' energy to break free from gravitons and, therefore, are unable radiate as detectable 'visible' or 'invisible' light. Therefore, whilst light is detectable (and, possibly, visible) at *c*, the asserted speed of light, within this new theory it is proposed that light photons are bound to the GLEW particles and are therefore invisible / undetectable in that these do no radiate any form of EMR. Such undetectable light and the associated energy radiated and sustaining GLEW stream velocity is proposed as the constituents of dark matter concentrations and dark energy radiation. This is due to an infinite number of GLEW streams travelling in infinite planes and dimensions (further to the law of the conservation of angular momentum).

There are six key concepts central to the following discussion:

1. Gravity travels as streams of particles in all conceivable directions: there is no such thing as 'up' or 'down' for gravity. Such streams vary in their velocity (speed and direction);
2. Dark matter and dark energy are composed of *Gravity-Light Energized Waves* (GLEW) streams;

41

3. Concentrations of GLEW streams around areas of greater mass and associated increases in the curvature of the space-time (STC) fabric form *dark matter*

4. The velocity and spin of the GLEW particles within each GLEW stream varies, thereby producing *dark energy*. That is, *dark matter* consists of higher concentrations of GLEW streams, and, as a consequence, the levels of dark energy will be much higher than areas of the universe (multiverse) where there are lower GLEW stream concentrations;

5. Gravity, in the form of GLEW streams, causes space-time curvature before the release of photons as detectable light.

6. The faster the GLEW streams travel in relation to maintaining the balance, direction and movement of cosmological bodies, the greater will be the space-time curvature created and followed.

7. All objects within the Universe and multiverse travel in the direction of least resistance, including living and non-living objects on Earth. That is, when there is imbalance or unbalance, GLEW (hereafter known as GLEW: abbreviated form of gravity-light order streams) will seek to create balance by moving an object in the direction and velocity of least resistance so that balance may be maintained as quickly as possible.

In summary, the key principles of GLEW theory are that:

1. Gravitons and photons travel as duality, co-joined streams throughout the Universe and Multiverse. (These are labelled as *Gravity-Light Energized Wave* streams – abbreviated to GLEW: pronounced glue)

2. Dark matter consists of an immeasurable and undetectable number of GLEW streams.

3. GLEW streams travel at variable velocities as means of sustaining balance and order within and between universes (via black holes).

4. GLEW streams have a velocity and momentum greater than the speed at which light energy is released: that is, a velocity greater than c as the asserted speed of *detectable* light.

5. Dark energy consists of energy released from and perpetuating the velocity of GLEW streams.

6. When GLEW streams decelerate, they will reach a velocity at which photon energy is released as detectable EMR and / or visible light. When GLEW streams decelerate to the point where photons are travelling at a velocity that light becomes detectable, this is the point at which the *Gravity-Light Acceleration-Related Energy (GLARE)* threshold velocity is achieved. Therefore, dark matter and dark energy is transformed to light energy which is detectable (and, in some cases, visible).

7. Black holes act as gateways, or pores, between universes within the Multiverse. If we use an analogy of a refrigeration unit, where heat exchangers ensure a balance of hot and cold EMR, the movement of dark matter and dark energy through black holes acts as a thermodynamic exchange system between universes, ensuring the balance of cosmological thermodynamics. As such. black holes act as thermodynamic regulation gateways (abbreviated to TRG) as a means of creating thermal equilibrium between three or more universes.

8. To ensure thermodynamic balance, each universe is linked to several other universes by numerous black holes acting as TRGs. Therefore, universes are not isolated systems: dark energy moves both into and out of a universe through TRGs but in equilibrium as a means of conserving energy within a given universe.

9. Black holes are akin to the swirling whirlpools or the eddies of a river, creating ripples that we generate gravitational waves in the form of GLEW streams of varying velocities.

10. There are different sizes of black holes, each of which has a varying event horizon. Each black hole is a tear in the fabric between universes.

11. The size of an individual black hole and its event horizon varies with the number of GLEW streams passing through it. Depending upon the number of GLEW streams, the black hole and its event horizon can be larger on one side than on the other.

12. The formulae for calculating the velocity of GLEW streams as they move through the universe, and between universes through black holes, is proposed as two formulas:

$$\mathbf{E}^{\mathrm{GLEW}} = \frac{\mathbf{v}^{\mathrm{GLEW}}\mathbf{v}^{\mathrm{m}}\mathbf{p}\mathbf{V}^{0}}{\mathbf{G}}$$

$$\mathbf{v}^{\text{GLEW}} = \frac{\mathbf{E}^{\text{GLEW}}\mathbf{v}^{\text{m}}\mathbf{p}\mathbf{V}^{\mathbf{0}}}{\mathbf{G}}$$

These two formulae are unravelled and discussed below.

Black holes maintain thermodynamic balance and order within the Multiverse

This novel theory of black holes as TRGs and of gravitons as the key transporter of photons within GLEW streams is an attempt to present a universal theory which successfully unifies theory, general relativity, 'string' theories, the 'symmetry' of the Universe, the Conservation of Angular Momentum law, and universal gravitation law. Such unity is explained by the presence and mechanics of the smallest possible particle but ones that we shall never be able to detect – the particles which combine to generate gravity.

All constituent universes within the Multiverse share a need to achieve and sustain balance and order. However, the movement and actions of other objects and matter creates imbalance and disorder. As a result, there is a need for simple, fast-acting processes which create the conditions for near-constant balance and stability. Our universe, other universes and our multiverse maintain balance (order) and minimise imbalance (disorder) through the combined mechanics and influence of gravity, light and thermodynamic-centric particles.

Central to this multiverse balance-imbalance theory are two key assertions. The first is that at the speed of detectable light, light and gravity travel together as duality streams. Travelling together enables the translation of *dark energy* (when the GLEW streams are travelling at a speed where photons cannot be detected but are still releasing dark energy) to *light energy (EMR)* as the streams decelerate to the accepted speed of light. Therefore, photons remain 'dark' and undetectable when they are travelling faster than the current accepted *mathematical* speed of light. In other words, when gravity and light travel at a velocity greater than the *mathematical* speed of light, they are travelling too fast for the accompanying photon(s) to generate and release detectable light.

Although the speed of GLEW is, in the main, greater than the known speed of detectable light, a key tenet of this theory is that the speed of gravity streams is variable. When the duality-stream decelerates by just a single unit (not defined within this current version of the theory), such as when it encounters an object with mass (including gaseous atmospheres), photons are released as detectable light and heat (transforming from dark matter and dark energy to light matter and light energy).

Black Holes as Thermodynamic Regulation Gateways (TRGs)

The four *laws of thermodynamics* define fundamental physical quantities (temperature, energy, and entropy) that characterize thermodynamic systems at thermal equilibrium. The laws describe how these quantities energize and thwart a range of phenomena. The zeroth law of thermodynamics states that if two systems are in thermal equilibrium with a third system, they are in thermal equilibrium with each other. This law helps define the concept of temperature. It is, therefore, posited that black holes act as thermal regulation gateways between one universe and at least two other universes as a means of attaining thermal equilibrium between the three universes (Zeroth law).

The first law of thermodynamics affirms that when energy passes, as work, as heat, or with matter, into or out from a system, the system's internal energy changes in accord with the law of conservation of energy. That is, this first law, also known as Law of Conservation of Energy, states that energy cannot be created or destroyed in an isolated system. Therefore, it is argued that black holes act as pores known as thermodynamic regulation gateways (TRG), on the understanding that a universe is not an isolated system. This means that energy moves both into and out of a universe but in equilibrium as a means of conserving energy within a given universe (as per the First Law).

Upon considering the second law of thermodynamics, within a natural thermodynamic process, the sum of the entropies of the interacting thermodynamic systems increases. In other words, the second law of thermodynamics states that the entropy of any isolated system always increases. This helps to understand black holes, as TRGs, as this means that the Universe (that we inhabit) and a multiverse are not isolated systems, in that each universe is linked to at least several other universes. Therefore, entropy may increase but it is balanced by the loss and arrival of new matter via the TRG.

Finally, it is stated within the third law of thermodynamics that the entropy of a system approaches a constant value as the temperature approaches absolute zero. The third law of thermodynamics states that the entropy of a system approaches a constant value as the temperature approaches absolute zero. The average temperature of the 'human' Universe is approximately 2.73 kelvins (−270.42 °C; −454.76 °F), based on measurements of cosmic microwave background radiation: that is, absolute zero! Black hole TRGs ensure that the Third Law is upheld and conserved, in that the entropy of a system at absolute zero is typically close to zero. Therefore, black holes ensure, via compliance with the Zeroth, First and Second law, that entropy is close to zero. That is, as black hole acts as TRGs, this means that universes are not isolated systems (First Law): instead they enable a thermal equilibrium between three or more universes (Zeroth Law) and balance entropy (Second Law) through the movement of matter to and from each universe as a means of maintaining absolute zero (Third Law).

In consequence, black holes, acting as TRGs, are thermodynamic regulators, with dark matter (DM) travelling in both directions within a single black hole gateway (pore). The balance of DM movement between two universes creates balance within each. The emergence of DM and dark energy (DE) from another universe into our Universe explains Hawking Radiation. In this case, Hawking Radiation would travel in both directions through each TRG: from the Universe into another universe, and, in exchange, from the same universe into the Universe. (N.B. The capital letter in the case of the Universe refers to the universe which humans currently inhabit!)

The boundaries of the Universe and all other universes within the Multiverse behave in the same way as we observe at the surface of water in, for example, a basin or a river. The Universe, as with all universes, is not a closed thermodynamic system, and would only be closed if it were not for the innumerable black holes in the fabric within the dark matter which is diffused within each universe. A key point is that the Hawking radiation and relativistic jets that are emitted from a black hole TRG consists of dark matter and dark energy that has travelled from an adjacent universe.

The images of a visible black hole accretion disc, within the Messier 87 galaxy (see, for example, https://www.eso.org/public/news/eso1907/: April 2019), suggest that black holes are TRGs for other universes, where the velocity of emitted Hawking radiation decelerates due to the presence of dust and gas creating an 'atmosphere' which causes $GLEW_v^0$ to decelerate to, at least, $GLEW_v^{-1}$, thereby emitting $GLARE^{+n.}$

GLEW streams consist of and create further dark energy and dark matter

At the heart of universal laws such as gravitation, the conservation of angular momentum, and thermodynamics, as well as theories of relativity, black holes and Hawking radiation, as we *perceive* and *understand* them to be, gravitons are the universal particle central to the gravity and the products of the momentum-velocity of gravity, especially dark matter and dark energy. Such a particle is present as gravity in such unimaginably large numbers that it creates the necessary levels of dark energy to travel faster than the asserted speed of detectable light and to be present in such large number as to be instantaneous in its presence and impact.

GLEW stream particles exist at instantaneous high densities and travel faster than light: by doing so, they *exist as both dark matter and dark energy*. The production and distribution of the density of dark matter is a means of creating balance and order within the surrounding dark energy. Therefore, gravity travels and instantaneously acts through its compositions as an innumerable series of GLEW streams. Although it is posited that GLEW streams travel faster than the known speed of light, this speed of GLEW streams is variable. When GLEW streams travel at their terminal velocity ($GLEW_v > c$), they are travelling too fast for the accompanying photon to release 'light' as detectable EMR. When the GLEW streams decelerate to the speed of light, such as when it encounters an object with mass (including gaseous atmospheres), the GLEW stream will be travelling just slow enough for photons to be released as detectable EMR in the form of visible or detectable light.

GLEW streams as the carrier of gravitons and photons

It is posited throughout this paper that GLEW streams consist of a series of undetectable individual GLEW particles. Each GLEW stream consists of graviton particles that are smaller than any other detectable particle, as they would need to be to be able to pass through all mass within the Multiverse and its constituent universes. Therefore, GLEW particles are proposed as the smallest of all sub-atomic particles, in order that they may pass through all mass without being absorbed. Conversely, it is proposed that the calculable influential processes of *gravity stems from the smallest possible, and as yet undetectable, sub-atomic particles.* These are GLEW particles, or gravitons and photons.

The attraction between GLEW particles leads to the formation of constant streams of GLEW particles. Such GLEW particles amass in 'strings' otherwise known as GLEW streams. As stated, central to this theory is the assertion that GLEW particles and GLEW streams travel faster than the known speed of light. *The speed of individual GLEW streams, although greater than the known speed of light, are variable.* When GLEW streams travel at their terminal velocity, they are travelling too fast for the accompanying photon(s) to release detectable light. When the GLEW stream decelerates from its terminal velocity, such as when it encounters mass-rich objects (including gaseous atmospheres), photons are released and detectable light. Therefore, I suggest that GLEW streams (as *gravity-light duality streams*) are the 'unknown' constituents of dark matter, in the form of energy-rich streams of gravity particles that are travelling too fast to release photons.

All structures with mass contain a greater concentration of GLEW particles due to the mass-rich accumulation and attraction of GLEW streams into and through an object's mass. This leads to an increased gravitational 'attraction' due to the higher concentration of GLEW streams and, by default, their individual GLEW particles. That is, the larger the object and the closer the distance to another object, the greater will be the movement of the GLEW streams between the two objects. This supports the principles of Newton's inverse square law of gravitation. Such an accumulation of GLEW particles would assist in explaining why some areas of the Earth have weaker or stronger gravitational force than other areas.

Further to mass-energy equivalence, the GLEW particles conserve gravitational force as they are constantly moving within GLEW streams or within structures in the form of 'building GLEW particles'. This constant moving takes two forms. The first is, as Richard Feynman described it, 'jiggling', which means that kinetic, heat and mass energy are produced and consumed. The generation and release of such energy results in the 'spin' and 'momentum' of all GLEW particles. GLEW particles are perpetually in motion even when they are part of a much larger structure, as they do not interact with any other quantum particle except photons. This perpetual spin results in each g-block attracting others in the proximity. That is, through the energy produced by the GLEW particles and the charge which each g-block possesses, there is a strong attraction between a single GLEW particle and the GLEW particles surrounding it. It also has a strong attraction for other GLEW particles as they are encountered.

A GLEW stream can exist in one of two forms: as a constant band of GLEW particles with photon(s) attached or as GLEW particles without a photon attached. As stated, within the GLEW stream. GLEW particles are constantly travelling both in the form of spinning and directional momentum / velocity. The GLEW particles do not all spin the same direction as each other: by spinning within different planes whilst tightly held in place, their directional spin balances each other and prevents imbalance within a GLEW stream. The speed of the spin and the speed of the GLEW stream momentum generates velocity greater than the known speed of light.

It is proposed that there is a constant stream of GLEW particles within a g-stream. The GLEW streams are analogous to the movement of water in a river, with each stream diverting and moving to fill gaps. The GLEW particles spin within the g-stream, which as a consequence causes each g-stream to spin. By spinning, the g-stream is constantly shifting the position of each of its GLEW particles in relation to each other, which in turn attracts other GLEW particles within a single g-stream and at the periphery of adjacent GLEW streams. This creates bonding within and between individual GLEW streams.

GLEW stream velocity variations are due to varying masses, atmospheres and dark matter concentration

Dark matter fills the spaces between stars, planets and other solid masses (comets, asteroids, etc.) within the universe. This dark matter consists of GLEW streams travelling in tandem with light at such a speed that the light remains bonded and unable to be released. Therefore, light, as photons, are not emitted and the medium is unseen as 'dark matter'. When a gravity-light-order (GLEW) duality stream moves from a less dense to a denser medium (such as the atmosphere of a planet, the planet structure, a natural satellite such as a moon, or a medium such as water), the gravity-light duality stream decelerates until it is reduced to the speed at which light is visible / detectable. Therefore, upon encountering a medium that is denser than dark matter, light is released as photons and is detectable in the form of electromagnetic radiation. For clarification, the 'medium' (see Figure 1 below) is any mass-rich object other than dark matter. The emitted light is either absorbed or reflected. For example, when a GLEW stream enters a planet's atmosphere the speed of the stream decelerates to a velocity where light will be released as photons.

Calculations explaining the variant velocity of GLEW streams

The formulae for calculating the energy and velocity of GLEW streams as they move through the universe, and between universes through black holes, are:

$$^{GLEW}E = \frac{^{GLEW}v^m vpV^0}{Gm^1m2/r^2}$$

$$^{GLEW}v = \frac{^{GLEW}Ev^mpV^0}{Gm^1m2/r^2}$$

Where:

GLEW$_E$ = Energy that is simultaneously generated and released as the means of maintaining the balance that we know as gravity (G) and to perpetuate the momentum of GLEW streams.

GLEWv = the velocity of the GLEW stream, with $\text{GLEW}_v{}^0$ being the velocity at which stability (balance and order), as a threshold, is achieved.

v^m = the velocity of a mass (object)

pV^0 = the thermodynamic constant as defined within the four laws of thermodynamics

$\overline{G\ m^1m^2\ /\ r^2}$ = the gravitational constant (Newton)

A key aspect of our understanding is that GLEW_E is simultaneously generated and released as the means of maintaining the balance that we know as gravity (G) and to perpetuate the momentum of GLEW streams. That is, GLEW_E is not stored but, instead, is released simultaneously upon its generation through GLEW_v.

This can be expressed as:

$$\text{GLEW}_E = \text{GLEW}_v = {}_m\text{GLEW}_v\ /\ G$$

The release of detectable light, including as visible light, is only possible when GLEW streams decelerate (from $\text{GLEW}_v{}^0$ to a minimum of $\text{GLEW}_v{}^{0-n}$). At the velocity of $\text{GLEW}_v{}^{0-n}$, the GLEW stream will be travelling at c. At this velocity, detectable and visible light is released as Gravity-Light Acceleration-Related Energy (or *GLARE*).

This can be expressed as follows:

Where light is neither detectable nor visible, then

$$\text{GLEW}_E{}^0 = \text{GLEW}_v{}^0 => c = \text{GLARE}^0$$

Where light is detectable, and potentially visible, then

$$\text{GLEW}_E{}^{-n} = \text{GLEW}_v{}^{-n} = c = \text{GLARE}^{+n}$$

That is, GLARE^{+n} = the radiation of detectable, and potentially visible, light. Therefore:

$$\frac{\text{GLEW}_E{}^n = \text{GLEW}_v{}^n = \text{GLARE}^{(+\sim)n}}{G}$$

That is:

1. No detectable light (GLARE^0) is generated when

$$\frac{\text{GLEW}_E{}^0 = \text{GLEW}_v{}^0 = \text{GLARE}^0}{G}$$

2. Detectable light (GLARE^{+n}) is generated when

$$\frac{\text{GLEW}_E{}^{-1} = \text{GLEW}_v{}^{-1} = \text{GLARE}^{+1}}{G}$$

3. GLARE is intensified with the further deceleration of GLEW streams, for example:

$$\frac{\text{GLEW}_E{}^{-2} = \text{GLEW}_v{}^{-2} = \text{GLARE}^{+2}}{G}$$

Based upon the universal gravitational constant (G), where the value of G is lower, then GLEW_E and GLEW_v will, conversely, be higher. Therefore, the presence of GLARE will be reduced or zero. Inversely, if G is higher, then GLEW_E and GLEW_v will be lower and the presence of GLARE will be higher.

Whilst detectable light is emitted from photons which are, individually, either reflected or absorbed, by contrast, the GLEW particles, as a GLEW stream, continues to travel through the mass of the planet and emerge from the other side (see Figures 1 and 2). As the GLEW stream is 'released from the denser medium (denser than dark matter), the velocity of the GLEW stream instantaneously reverts to $\text{GLEW}v^0$ (The velocity of GLEW streams through and as both dark matter and dark energy).

GLEW streams travel instantaneously in all directions around and through objects of different densities and mass. The reason that the GLEW streams approach, enter and exit from objects in an infinite number of different directions is due to attraction between the GLEW particles within the stream and the GLEW particles within the 'fixed' structure that the stream is travelling through. Clearly, for gravity to have a universal impact, the streams must enter and exit a structure in an infinite, inestimable number of directions. However, if GLEW streams are to inform our understanding of quantum mechanics and general relativity in terms of the curvature of the fabric of space, GLEW streams curve as they consist of and generate dark matter.

The curvature within the dark matter field is not due to the deceleration of the GLEW stream: a GLEW stream constantly travels at a speed greater than the speed at which light becomes detectable by available instruments. It is not within the remit of the present paper to propose what that speed might be, other than it has been expressed, for the sake of illustrating differing velocities, as $GLEWv^0$. The curvature of space is entirely due to the curvature of GLEW streams: they are the same entity in that the curvature of space is indicative of the curvature of an innumerable and unmeasurably large number of GLEW streams.

GLEW streams are integral to space-time curvature and black holes

In 2020, it was acknowledged that the GLEW theory, and the principle of GLARE, is in the very early stages of its development. There is room for further development and modification. However, through collaboration with other researchers, I believe that it has the potential to develop as a means of explaining and understanding gravity combines quantum mechanics, Einstein's two theories of relativity and thermodynamics within contained universes and the whole multiverse. Most importantly, it may be used to further inform our understanding of black holes and gravity. The curving of the space-time fabric of the universe is due to the curvature of the GLEW streams, and changes in the length of the GLEW streams.

It is proposed that whilst the speed of detectable light remains constant (as c), the speed of GLEW streams varies (expressed as $GLEWv^n$). However, the velocity of the GLEW streams ($GLEWv^n$) is never slower than but can be equal to the speed of detectable light (that is, $GLEW_v{}^n \geq c$). Where GLEW streams are travelling at a velocity greater than c, the structure of dark matter is gravity at its strongest ($GLEWv^0$), and therefore objects must travel faster in order to counteract the influences of gravitational streaming. As the GLEW streams (which have both mass and energy) encounter new mass-rich objects, such as stars and planets, the number of GLEW streams increases, moving towards the mass-rich object, thereby causing a curvature in the gravitational field.

Figure 2 *The changes in GLEW stream velocity (GLEW,^{n}) due to differing mass densities* (see Key in Figure 1)

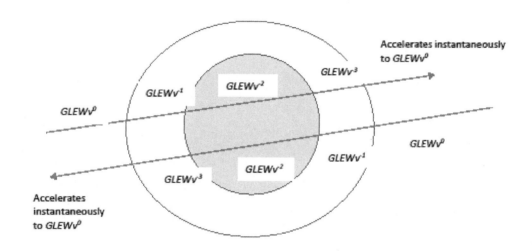

The increased density of GLEW streams attracts more GLEW streams, thereby leading to a greater detectable gravitational streaming (or pull / attraction). That is, that the greater the mass of GLEW streams, the greater the gravitational attract due to enhanced energy (on the basis of mass-energy equivalency). The acceleration and deceleration of GLEW streams results in the formation of gravitational waves, due to changes in velocity: this creates a ripple (or wave) in the movement of GLEW streams. This is due to individual and adjacent GLEW streams being able to expand and stretch, and to contract and shorten. This would lead to an overlap of gravitational influences and gravitational waves similar to the ripple effect when several stones are thrown into a lake in a fashion similar to 'Newton's rings'.

GLEW streams are the basis of the 'curvature of the dark matter within the balloon shape of the Universe that we inhabit. That is, light curves because the gravity curves: such a curvature in the light that travels from a star is, of course, only detectable on Earth when the light is released upon the entry of the g-stream into the atmosphere. Within our own solar system, the large gaseous planets, such as Jupiter and Saturn, have rocky centres with huge masses, surrounded by substantial gaseous atmospheres exert a large gravitational force. This is due to the large number of GLEW particles within the fixed structure of the mass of the planet itself and the atmosphere surrounding it. As GLEW streams travel as dark matter, the GLEW particles within the g-stream are attracted to the GLEW particles that form the structure (rock and atmosphere) of the planets. This causes a curvature in the path of the GLEW streams, which each react to the movement of the others. The duality between gravity and light would account for the curvature of light as it passes the, in this case, a planet or a star. Therefore, the curvature of space *is* the curve of GLEW streams as they are attracted towards GLEW streams that are exiting the planet or star. This space-time curvature will be based upon the mass of the object and explains where there are varying gravitational influences depending upon the concentration of GLEW particles within a particular medium, such as, for example, rocks.

A question for consideration: is our Universe contained within a black hole?

Within GLEW theory as it currently stands, GLEW particles form the composition and the energy that velocity of GLEW streams, an intriguing thought arises: that the universe we inhabit (and everything that constitutes that universe) is contained *inside a black hole*. That is, the reason that we believe that we have only found only one black hole (Cygnus X-1), at the time of writing, is that we are within Cygnus X-1 looking outwards to our own event horizon. Therefore, if we presume that we exist *within* a closed thermodynamic system where the expansion of the universe (in the form of the galaxies moving further away from each other) is due to an expansion in the size of the black hole. If we exist and are travelling *within* a black hole, we would, therefore, if we could see a black hole, not be looking into the black hole. Instead, we would be looking outwards to the event horizon. As the event horizon expands, due to a loss of mass (as proposed within the theory of Hawking heat radiation), the event horizon will become less visible. If our own galaxy, together with all other known galaxies, exist and travel within a black hole, this, in itself, may be regarded as a universe with an interior that is constantly expanding. This would, it is envisaged, be like the inside of an ever-expanding balloon that has expanded beyond the limits of our ability to see to the edge.

In addition, the presence and movement of our own and other galaxies within a black hole may well be the reason that dark matter is perceived as *dark*: that is, that the dark matter is the 'blackness' of the inside of a black hole. Therefore, if we are looking outwards towards the event horizon of the black hole that we are within, the analogy of a 'balloon' shape to the black hole would enable the curvature of space-time. As the event horizon expands, it is suggested that the mass of GLEW streams leaving the event horizon (i.e. the source of Hawking heat radiation) will eventually be greater than the mass of GLEW streams entering the black hole. This will lead to a point where the event horizon is no longer able to maintain its shape and will expand suddenly. This will result in a sudden expansion, leading to the reaction equivalent to the 'big bang': the event horizon will expand suddenly to the point where the contents of the black hole become enclosed within another, more expansive black hole.

As a black hole simultaneously loses and gains Hawking radiation, this takes the form of GLEW streams where gravity and light are travelling as GLEW streams. As the black hole interior expands, more GLEW particles (within associated GLEW streams) are absorbed into the black hole than are being released (known as Hawking radiation) from the black hole / universe that encloses it. As the black hole gains more gravitational mass, due to a greater influx of

GLEW streams than radiated GLEW streams, the black hole expands to the point where it merges with the surrounding universe that contains it. This expansion of numerous event horizons at different points in the history of the Universe (as a whole), would lead to universes combining. Therefore, each universe would once have been part of the contents of a previous black hole.

By drawing further upon the principles of black holes being TRGs, which emit Hawking radiation in both directions, if one accepts that our universe is actually the contents of a black hole and that we are looking not towards a black hole but indeed outwards from within a black hole, it can therefore be determined that each known universe is contained within a black hole. Each black hole, therefore, contains at least one universe. As stated previously, The recent images of a visible black hole accretion disc, within the Messier 87 galaxy, reveals that black holes are TRGs for other universes, where the velocity of emitted Hawking radiation decelerates due to the presence of dust and gas creating an 'atmosphere' which causes $GLEW_v^0$ to decelerate to, at least, $GLEW_v^{-1}$, thereby emitting $GLARE_v^{+n.}$

Therefore, let us presume that we are looking outwards towards the event horizon and in to another universe within another larger black hole that contains our own Universe / black hole. That is, each black hole is contained within a larger black hole, which is within an even larger black hole, and so on and so forth. The principles of Hawking radiation suggest that black holes have a thermodynamic instability in that mass is released from the event horizon as heat radiation as opposed to detectable light. Therefore, it is proposed that the heat radiation and thermodynamic instability is due to the movement of GLEW streams from one black hole, via the event horizon, to the larger, encompassing black hole through a form of gravitational partition. This accretion of GLEW streams (and the resultant movement of gravity-light-heat) results in a gradual expansion in the event horizon almost like one sees in the neck of a balloon when the air is released.

Eventually the balance of GLEW movement leads to an instability at the event horizon. This leads to the instability and expansion of the event horizon to the extent that the smaller universe within the black hole becomes subsumed in to the larger universe within another black hole.

Therefore, each singularity expansion leads to the formation of a new universe that is much larger than its predecessor due to the merging of two universes: a result of the expansion and ultimate instability of the smaller universe's event horizon. As such, it is proposed that each singularity expansion leads to the creation of larger, merged universes: this expansion (whether gradual over billions of years or sudden and 'explosive') is due to Hawking radiation being released to the extent that an event horizon expands suddenly, leading to the immense sudden release and potential redirection of GLEW streams. If this is a sudden and violent episode in the expansion and merging of universes, this would have a catastrophic impact upon objects within both the smaller and larger merged universes.

The least catastrophic event would the gradual expansion of the event horizon, due to the reciprocal movement of GLEW streams across the event horizon, leading to the merging of the two black holes, i.e. the universe within the inner black hole gradually becomes part of the larger black hole due to the shell of the smaller black hole gradually receding. Therefore, the event horizon gradually expands to the point where it is no longer discernible and the universe it held in place becomes indistinguishable from the universe surrounding it.

For such a theory to be plausible, a number of key assumptions need to be considered. The first is that each universe (and the black hole within which each universe is contained) is a self-contained thermodynamic entity called a closed thermodynamic system (CTS). Each CTS universe is ever-expanding within a rotating black hole with a shell. Each CTS expands due to the movement of GLEW streams. This expansion enables and is vital to the gravitational impact of the movement of the GLEW particles in that the GLEW streams are perpetually moving within the closed unit of the black hole universe. This leads to the rotation and curvature of GLEW streams within orbits that would explain the movement of planets within the various solar systems and the curvature of space fabric, as well as the expansion of the universe. It is important to reiterate here that each universe is a closed thermodynamic system within an infinite series of systems.

The event horizon enables the movement of GLEW streams from the larger (surrounding) universe whilst GLEW streams (gravity and heat) radiate from the universe (via the event horizon) as the equivalent of Hawking radiation. As long as a balance is made between the movement of GLEW in and out – in terms of mass and energy – the gravitational balance with remain constant. The expansion of the universe (within the black hole) will lead to a point where the capacity of the universe is such that its shell can no longer contain all of the GLEW streams and their member GLEW particles. In consequence, radiation of GLEW streams out of the universe will be greater than the influx of such streams. This will lead to an expansion of the event horizon and the gradual loss of the outer shell of the black hole.

Each energy-rich, explosive expansion of a black hole would lead to a reaction similar to that of a singularity expanding to fill the known space within the Universe: that is, that one universe combines with another universe to become more mass rich. In turn, each becomes part of the same, but larger, closed thermodynamic system. Each of these 'singularity-equivalent' event horizon expansions are asserted as being the equivalent of a 'big bang' reaction, and involves both sudden expansion of the galaxies from one universe into the galaxies of another larger universe. As part of this expansion process, there is an increase in the velocity of the GLEW streams (to a velocity of $>GLEW_v{}^{+1}$) which should be detectable as gravitational super-waves. These are of a greater magnitude than detectable gravitational waves, the latter travelling at gv^0. Each gravitational super-wave represents the sudden expansion of a black hole event horizon, and the sudden, instantaneous rush of GLEW streams from one black hole in to a larger black hole. Both gravitational waves and gravitational super-waves are indicative of sudden changes in the velocity of GLEW streams, both, respectively, of the 'big bang' expansion of black hole event horizons and changes due to mass / density changes. These will be detectable in similar ways to one detects the ripples on the surface of a pond. Given that there may be any number of event horizons expanding at any one time, there will be a series of overlapping ripples of gravitational super-waves and waves (again, due to changes in GLEW stream velocity and associated oscillations).

A further key principle to take into account if such a model of universal gravitation is to be viable is that the GLEW streams are able to contract and stretch in different parts of the universe. Such contraction and expansion would enable the predictability of the orbits of mass-rich objects around each other, including planets around stars, stars around the spiral within their own galaxy, and the orbits of comets and constituents of asteroid showers. Such contraction and expansion of the GLEW particles within GLEW streams would explain why planets and stars have aphelia and perihelia. This change in length enables changes in distances between objects whilst keeping them bound within a system which has an almost 'clockwork' predictability.

The third principle is that there may be a different form of the black hole within the proposed form of black hole that forms the CTS that is a universe. This is not so much a hole leading to a universe but a series of areas of such GLEW particle concentration, due to the extremely high number of GLEW streams being pulled through due to mass-gravity attraction, that is necessary to ensure the sustained predictability of the orbits of mass-rich objects. In essence, these are dark voids within the universe that have strong gravitational fields and radiate large amounts of heat but do not release light due to the speed of the GLEW streams through the dark void (i.e. speeds of $GLEW_v{}^0$ and $GLEW_v{}^{+1}$). These dark voids of the highest concentration of GLEW streams act like tension points (almost like the tension points where several knots meet in a cargo net) around which the movement of GLEW streams takes place to enable orbital predictability within an expanding universe.

The evolution of GLEW and its emergence as UE-GLEW Theory

I concluded the previous book by stating that a number of concepts central to the discussed GLEW Theory need to be revisited if we are to be more able to understand the systems that maintain balance within our own Universe and the multiverse. This includes the understanding that dark matter, including non-detectable light, is capable at travelling at greater the asserted velocity of detectable light. For the model to work and the theory to be viable, dark matter needs to be capable of moving at faster than the speed of light. The speed of dark matter (DM) is such that, whilst dark energy (DE) may be generated and released, the velocity of DM is such that 'light' as a form of DE is generated but cannot be released at normal DM velocity. Therefore, on that basis, we need to rethink the 'speed of light' as being the 'speed at which light is detectable'. I also stated that, in summary, for the discussed theory of GLEW to be accepted as a basis for enhancing our understanding of gravity, dark matter and black holes, we need to rethink our entrenched thinking regarding the speed of light, the mechanics of gravity and the forms that black holes take if the particles that constitute and govern them are of the smallest possible form: that is, GLEW particles.

The UE-GLEW theory presented in the current volume is a response to the recognitions that, in 2020, the GLEW theory is in its infancy and would be enriched by further details and ideas as part of its evolution. A key element is determining (or, at the very least) suggesting the range of the actual speeds at which GLEW particles and streams need to travel in order to generate dark matter and dark energy. It may be that GLEW particles have a *mass* that, individually, has a very limited gravitational influence in that only a small amount of energy is released by a single GLEW particle. However, if all matter is composed of GLEW at the very core of its structure, there will be implications in terms of the mass-energy equivalency. In addition, this may explain why GLEW stream velocity causes planets and stars to rotate and develop a typical spherical or oblate spherical shape based upon the principle that gravity pulls matter together but rotation throws it apart.

If the scientific world is accepting of the view that gravity is mass and mass is gravity, in that all GLEW particles have a mass-energy equivalency, and that GLEW particles travel faster than the asserted speed of light, a speed at which light as an energy form is neither released or detectable, then we are far better placed to unite quantum and general relativity theories. As with the similar proposal of related theories of gravity, only time and space will reveal.

CHAPTER 4

Backwards through Time as a basis for Change!

As stated in the previous chapter, I concluded my 2020 book by asserting that a number of concepts central to the discussed GLEW Theory need to be revisited if we are to be more able to understand the systems that maintain balance within our own Universe and the multiverse. This includes the understanding that dark matter, including non-detectable light, is capable at travelling at greater the asserted velocity of detectable light. For the model to work and the theory to be viable, dark matter needs to be capable of moving at faster than the speed of light. The speed of dark matter (DM) is such that, whilst dark energy (DE) may be generated and released, the velocity of DM is such that 'light' as a form of DE is generated but cannot be released at normal DM velocity. Therefore, on that basis, we need to rethink the 'speed of light' as being the 'speed at which light is detectable'. I also stated that, in summary, for the discussed theory of GLEW to be accepted as a basis for enhancing our understanding of gravity, dark matter and black holes, we need to rethink our entrenched thinking regarding the speed of light, the mechanics of gravity and the forms that black holes take if the particles that constitute and govern them are of the smallest possible form: that is, GLEW particles.

The UE-GLEW theory presented in the current volume is a response to the recognitions that, in 2020, the GLEW theory is in its infancy and would be enriched by further details and ideas as part of its evolution. A key element is determining (or, at the very least) suggesting the range of the actual speeds at which GLEW particles and streams need to travel in order to generate dark matter and dark energy. It may be that GLEW particles have a *mass* that, individually, has a very limited gravitational influence in that only a small amount of energy is released by a single GLEW particle. However, if all matter is composed of GLEW at the very core of its structure, there will be implications in terms of the mass-energy equivalency. In addition, this may explain why GLEW stream velocity causes planets and stars to rotate and develop a typical spherical or oblate spherical shape based upon the principle that gravity pulls matter together but rotation throws it apart.

If the scientific world is accepting of the view that gravity is mass and mass is gravity, in that all GLEW particles have a mass-energy equivalency, and that GLEW particles travel faster than the asserted speed of light, a speed at which light as an energy form is neither released or detectable, then we are far better placed to unite quantum and general relativity theories. As with the similar proposal of related theories of gravity, only time and space will reveal.

Although initially treated with doubt, scepticism, and, in some cases, ridicule, from many in the scientific community, Einstein's works gradually came to be recognised as significant advancements. In 1905, after publishing his work on special relativity, Einstein began working to extend the theory to gravitational fields; he then published a paper on general relativity in 1916, introducing his theory of gravitation. He continued to deal with problems of statistical mechanics and quantum theory, which led to his explanations of particle theory and the motion of molecules. He also investigated the thermal properties of light and the quantum theory of radiation, the basis of laser, which laid the foundation of the photon theory of light. In 1917, he applied the general theory of relativity to model the structure of the universe. From 1933 until his death in 1955, Einstein tried to develop a unified field theory of quantum physics and gravity.

One of his four famous 1905 papers postulated that energy is exchanged only in discrete amounts (quanta) (Ashok, 2003). This idea was pivotal to the early development of quantum theory (Spielberg and Byron, 1995). The aforementioned paper, "On the Electrodynamics of Moving Bodies"(published 26th September 1905), reconciled conflicts between Maxwell's equations (the laws of electricity and magnetism) and the laws of Newtonian mechanics by introducing changes to the laws of mechanics (Folsing, 1997). Observationally, the effects of these changes are most apparent at high speeds (where objects are moving at speeds close to the speed of light). The theory developed in this paper later became known as Einstein's special theory of relativity.

The General Theory of Relativity (GTR) is a theory of gravitation that was developed by Einstein between 1907 and 1915. According to general relativity, the observed gravitational attraction between masses results from the warping of space and time by those masses. General relativity has developed into an essential tool in modern astrophysics. It provides the foundation for the current understanding of black holes, regions of space where gravitational attraction is so strong that not even light can escape. In 1907, Einstein published "On the Relativity Principle and the Conclusions Drawn from It", he argued that free fall is really inertial motion. This argument is called the equivalence principle. In the same article, Einstein also predicted the phenomena of gravitational time dilation, gravitational redshift and deflection of light (Pais, 1982; Stachel et al., 2008). In the same year, Einstein proposed a model of matter where each atom in a lattice structure is an independent harmonic oscillator. In the Einstein model, each atom oscillates independently—a series of equally spaced quantized states for each oscillator.

Einstein was aware that getting the frequency of the actual oscillations would be difficult, but he nevertheless proposed this theory because it was a particularly clear demonstration that quantum mechanics could solve the specific heat problem in classical mechanics. In 1909, Einstein demonstrated that Max Planck's energy quanta must have well-defined momenta and act in some respects as independent, point-like particles. This paper introduced the *photon* concept (although, it is noted that the term '*photon*' was introduced later, by Gilbert N. Lewis in 1926) and inspired the notion of wave–particle duality in quantum mechanics. Einstein saw this wave–particle duality in radiation as concrete evidence for his conviction that physics needed a new, unified foundation. In 1911, Einstein published another article "On the Influence of Gravitation on the Propagation of Light" expanding on the 1907 article, in which he estimated the amount of deflection of light by massive bodies. Thus, the theoretical prediction of general relativity could, for the first time, be tested experimentally (Pais, 1982).

Furthermore, in 1916 and 1918, Einstein predicted and discussed gravitational waves: these are defined as ripples in the curvature of spacetime which propagate as waves, traveling outward from the source, transporting energy as gravitational radiation (Einstein, 1916,1918). He stated that gravitational waves cannot exist in the Newtonian theory of gravitation, as Newton postulated that the physical interactions of gravity propagate at infinite speed.

In 1917, Einstein applied GTR to the structure of the universe as a whole, predicting a Universe that was dynamic and expanding (Einstein, 1917). An interesting forerunner for TRGs is Einstein's 1931 theory, written in a previously unknown or overlooked manuscript, exploring a model of the expanding universe in which the density of matter remains constant due to a continuous creation of matter, a process he associated with the cosmological constant (Nussbaumer, 2014; O'Raifeartaigh et al., 2014). As he stated in the paper, "In what follows, I would like to draw attention to a solution to equation (1) that can account for Hubbel's [*sic*] facts, and in which the density is constant over time" ... "If one considers a physically bounded volume, particles of matter will be continually leaving it. For the density to remain constant, new particles of matter must be continually formed in the volume from space." This, in effect, was Einstein's argument for a steady-state universe (O'Raifeartaigh et al., 2014). However, whilst Einstein played a major role in developing quantum theory, beginning with his 1905 paper on the photoelectric effect, he was sceptical as to whether the randomness of quantum mechanics was fundamental rather than the result of determinism (Pais, 1979).

Indeed, Einstein believed that while the correctness of quantum mechanics was not in doubt, he still regarded it as, in some way, incomplete. That being said, Einstein's ideas about the existence of *entangled quantum states* led to the Einstein-Podolsky-Rosen (EPR) paper in 1935, which is still considered as central to the development of quantum information theory (Fine, 2017). Following his research on general relativity, Einstein entered into a series of attempts to generalize his geometric theory of gravitation to include electromagnetism as another aspect of a single entity. In 1950, he described his "unified field theory" in a *Scientific American* article titled "On the Generalized Theory of Gravitation" (Einstein, 1950).

A fundamental lesson of general relativity is that there is no fixed spacetime background, as found in Newtonian mechanics and special relativity; the spacetime geometry is dynamic. While easy to grasp in principle, this is the hardest idea to understand about general relativity, and its consequences are profound and not fully explored, even at the classical level. To a certain extent, general relativity can be seen to be a relational theory, in which the only physically relevant information is the relationship between different events in space-time (Smolin, 2001).
On the other hand, quantum mechanics has depended since its inception on a fixed background (non-dynamic) structure. In the case of quantum mechanics, it is time that is given and not dynamic, just as in Newtonian classical mechanics.

String theory can be seen as a generalization of quantum field theory where instead of point particles, string-like objects propagate in a fixed spacetime background, although the interactions among closed strings give rise to space-time in a dynamical way. it was soon discovered that the string spectrum contains the graviton.

One of the conceptual difficulties of combining quantum mechanics with general relativity arises from the contrasting role of time within these two frameworks. In quantum theories time acts as an independent background through which states evolve, leading to the generation of infinitesimal translations of quantum states through time (Sakurai and *Napolitano, 2010)*. In contrast, general relativity treats time as a dynamical variable which interacts directly with matter, implying that this removes any possibility of employing a notion of time similar to that utilised within quantum theory (Novello and Bergliaffa, 2003). As a consequence of being unwilling to reconsider how we define key aspects of GTR and quantum theory, we have been stymied in that we still do not have a complete and consistent quantum theory of gravity. This is the first challenge, although it is acknowledged that, as yet, there is no way to put quantum gravity predictions to experimental tests, although there is hope for this to change as future data from cosmological observations and particle physics experiments becomes available (Ashtekar, 2007; Schwarz, 2007). An important aspect of quantum gravity relates to the question of coupling of spin and spacetime. While spin and spacetime are expected to be coupled, the precise nature of this coupling is currently unknown (Yuri, 2001). In particular, it is not known how quantum spin sources gravity, nor what is the correct characterization of the spacetime of a single spin-half particle.

The General Theory of Relativity (GTR: Einstein, 1915) is regarded as the current description of gravitation in modern physics. The GTR was presented as a geometric theory of gravitation. Einstein utilised tensor calculus as the mathematical tool for formulating GTR. Within the Special Theory of Relativity (1905), Einstein presented mass as a form of energy, and energy as liberated mass. He presented space and time, within four dimensions, as dependent upon relative motion. With GTR, Einstein asserted that gravity is a function of mass and that gravity affected light in the same way as it had an impact upon mass particles. Light, therefore, did not always go 'straight'.

GTR presented a novel and unified description of gravity as a geometric property of space and time, or spacetime: specifically, that the *curvature of spacetime* is directly related to the energy and momentum of mass and radiation. Therefore, whereas mass and momentum had previously been considered as independent of space and time, GTR brought the four factors together.

GTR has had important astrophysical implications and led to key outcomes. For example, it *implied* the existence of black holes, defined as *regions* of space in which space and time are distorted in such a way that nothing, not even light, can escape: this was implied as an end-state for massive stars. In addition, there is ample evidence that the intense radiation emitted by certain kinds of astronomical objects is due to black holes. For example, microquasars and active galactic nuclei appear to result from the presence of stellar black holes and supermassive black holes, respectively. The bending of light by gravity can lead to the phenomenon of gravitational lensing, in which multiple images of the same astronomical object are visible. GTR also predicted the existence of gravitational waves, which have since been observed directly by the physics collaboration LIGO. In addition, general relativity is the basis of current cosmological models of a consistently expanding universe.

As an outcome of GTR, physicists felt that they had an increasing understanding of black holes, with quasars being identified as a manifestation of BH, in the form of collapsed stars. (Within GLEW Theory, black holes are not defined as holes but as black voids or black regions in space). A key strength of GTR, however, is its simplicity and symmetry, as well as the importance of incorporating invariance and unification with a logical consistency. Since the publication of GTR, a number of unanswered questions remain, especially the fundamental question of how general relativity can be reconciled with the laws of quantum physics to produce a *self-consistent theory of quantum gravity*. GLEW is presented as a potential answer within the realms of relativistic and quantum cosmology.

However, it has been stated that the application of GTR does not reduce the difficulty of finding exact solutions to ongoing cosmological questions such as the existence of naked singularities within black holes (as defined as collapsed stars and not as true black holes: see Chapter 2) and the movement of matter in relation to the detectable-visible asserted speed of light. Light follows a *light-like* or *null geodesic* trajectory—a generalization of the straight lines along which light travels in classical physics.

GLEW Theory is a potential theoretical response to an even larger question, and that is the physics of the earliest universe, prior to the inflationary phase and close to where the classical models predict the big bang singularity. Indeed, there is a general set of laws called the 'black hole mechanics'. These are analogous to the laws of thermodynamics. For instance, according to the second law of black hole mechanics, the area of the event horizon of a general black hole will never decrease with time, in the same vein as the entropy of a thermodynamic system. This limits the energy that can be extracted by classical means from a rotating black hole (e.g. by the Penrose process). However, it may be the laws of black hole mechanics are, in fact, a subset of the laws of thermodynamics, and that the black hole area is proportional to its entropy. This, if so, would require a modification of the original laws of black hole mechanics. That is, for example, as the second law of black hole mechanics becomes part of the second law of thermodynamics, it is possible for black hole area to decrease. However, this may only happen as long as other processes lead to overall increases in entropy. As thermodynamical objects with non-zero temperature, black holes should emit thermal radiation. Semi-classical calculations indicate that indeed they do, with the surface gravity playing the role of temperature in Planck's law. This radiation has been known, since 1974, as Hawking radiation.

Whilst our current conceptualization and understanding of gravity is based on GTR, within classical physics, the description of gravity has been considered as incomplete. For example, if one considers the gravitational field of a black hole as implied within the GTR, physical phenomena, such as the space-time curvature, may diverge at the centre of the black hole. This would result in an inconsistency within GTR, resulting in the need for a theory that goes beyond general relativity.

If we are to move forward and gain a more informed understanding of these quantum effects, a theory of quantum gravity is needed. Such a theory should allow the description to be extended closer to the center and might even allow an understanding of physics at the center of a black hole. On more formal grounds one can argue that a classical system cannot consistently be coupled to a quantum one (Feynman et al., 1995; Wald, 1984). It is important that any new theory of quantum gravity should provide potential, if not absolute, explanations of how gravity and light interact to influence the momentum of massive bodies across the Multiverse, as well as explaining the structure and function of black holes, and how mass and energy is regulated, renewed and redistributed across individual universes.

Of course, there are still many unanswered questions. Those that readily spring to mind are:

- What is the role of UE in maintaining an object's elliptical path, that are predictable and responsive to change whilst the universe is expanding

- With regards to the rotation of BHs, in relation to their purpose, is there a minimum size that a black hole can be?

- What is BH plasma, and how does this relate to UE?

- How can we determine if we are looking into a BH-TRG or out of a BH-TRG? That is, are we inside the room or outside the room? (How, if at all, can this be explained by GLEW / GLUE theory?)

- How do we explain tidal action if gravity is an effect-response?

A number of concepts central to the discussed GLEW Theory need to be revisited if we are to be more able to understand the systems that maintain balance within our own Universe and the multiverse. This includes the understanding that dark matter, including non-detectable light, is capable at travelling at greater the asserted velocity of detectable light. For the model to work and the theory to be viable, dark matter needs to be capable of moving at faster than the speed of light. The speed of dark matter (DM) is such that, whilst dark energy (DE) may be generated and released, the velocity of DM is such that 'light' as a form of DE is generated but cannot be released at normal DM velocity. Therefore, on that basis, we need to rethink the 'speed of light' as being the 'speed at which light is detectable'.

In summary, for the presented and discussed theory of UE-GLEW to be accepted as a basis for enhancing our understanding of gravity, dark matter and black holes, we need to rethink our entrenched thinking regarding the speed of light, the mechanics of gravity and the forms that black holes take if the particles that constitute and govern them are of the smallest possible form: that is, GLEW particles. Clearly the GLEW theory is in its infancy and will be enriched by further details and ideas as part of its evolution. A key element is determining (or, at the very least) suggesting the range of the actual speeds at which GLEW particles and streams need to travel in order to generate dark matter and dark energy. It may be that GLEW particles have a *mass* that, individually, has a very limited gravitational influence in that only a small amount of energy is released by a single GLEW particle. However, if all matter is composed of GLEW at the very core of its structure, there will be implications in terms of the mass-energy equivalency. In addition, this may explain why GLEW stream velocity causes planets and stars to rotate and develop a typical spherical or oblate spherical shape based upon the principle that gravity pulls matter together but rotation throws it apart.

If the scientific world is accepting of the view that gravity is mass and mass is gravity, in that all GLEW particles have a mass-energy equivalency, and that GLEW particles travel faster than the asserted speed of light, a speed at which light as an energy form is neither released or detectable, then we are far better placed to unite quantum and general relativity theories. As with the similar proposal of related theories of gravity, only change and space will reveal if this possible.

REFERENCES and BIBLIOGRAPHY

Brooks, M. (2005) 13 Things That Do Not Make Sense. **New Scientist** Issue 2491. [Last accessed 14th October 2021]

Desai, S. and Popławski, N.J. (2016) Non-parametric reconstruction of an inflaton potential from Einstein–Cartan–Sciama–Kibble gravity with particle production. **Physics Letters B**, 755: 183–189.

Easson, D.A., and Brandenberger, R.H. (2001) Universe generation from black hole interiors. **Journal of High Energy Physics**, 06, 024.

Einstein, A. (1903) Kinetic Theory of Thermal Equilibrium and the Second Law of Thermodynamics, **Annalen der Physik**, 9, 417 – 433.

Einstein, A. (1905a) On a Heuristic Point of View Concerning the Production and Transformation of Light. **Annalen der Physik**, 17, 132 - 148.

Einstein, A. (1905b) On the Electrodynamics of Moving Bodies. **Annalen der Physik**, 17, 891 - 921.

Einstein, A. (1905c) Does the Inertia of a Body Depend upon its Energy Content? **Annalen der Physik**, 18, 639 - 641.

Einstein, A. (1916 / 2010 translation). **Relativity: The Special and the General Theory.** Mansfield, CT: Martino Publishing.

Einstein, A. (1950) On the Generalized Theory of Gravitation. **Scientific American**, (4): 13–17.

Eriksen, H.K., Hansen, F.K., Banday, A.J., Górski, K.M. and Lilje, P.B. (2004) Asymmetries in the Cosmic Microwave Background Anisotropy Field. **The Astrophysical Journal**, 605 (1): 14–20.

Fixsen, D. J. (2009. The Temperature of the Cosmic Microwave Background. **The Astrophysical Journal**, 707(2): 916–920.

Frolov, V.P., Markov, M.A. and Mukhanov, V.F. (1989) Through a black hole into a new universe? **Physics Letters B,** 216: 272.

Hanson, N.R. (1958) **Patterns of Discovery: An Inquiry into the Conceptual Foundations of Science.** Cambridge: Cambridge University Press.

Hanson, N.R. (1969) **Perception and Discovery: An Introduction to Scientific Inquiry**. US: Freeman, Cooper and Co.

Hanson, N.R. (1971) **Observation and Explanation: A Guide to Philosophy of Science.** (Harper Essays in Philosophy) London: Harper and Row.

Hawking, S.W. (1994). **Black holes and baby universes and other essays.** London: Bantam Books.

Hawking, S.W. (2008) **The Theory of Everything: The Origin and Fate of the Universe.** Mumbai, India: Jaico.

Hawking, S.W and Ellis, G.F.R. (1973) **The Large Scale Structure of Space-Time.** Cambridge, UK: Cambridge University Press. ISBN 978-0-521-09906-6.

Hawking, S.W. and Israel, W. (1989) **Three Hundred Years of Gravitation.** Cambridge, UK; Cambridge University Press.

Hawking, S.W. and Penrose, R. (1996) **The Nature of Space and Time.** Princeton, NJ: Princeton University Press.

Isaacson, W. (2007) **Einstein: His Life and Universe.** London: Simon and Schuster.

Isham, C.J. (1993) Canonical Quantum Gravity and the Problem of Time. Integrable Systems, Quantum Groups, and Quantum Field Theories. **NATO ASI Series**. Springer, Dordrecht. pp. 157–287.

Kanigel, R. (1991) **The Man Who Knew Infinity: A Life of the Genius Ramanujan.** London: Abacus.

Massey R. (lead author, plus19 other authors) (2007) Dark matter maps reveal cosmic scaffolding. **Nature:** see https://www.nature.com/articles/nature05497 [accessed 9th November 2021]

Mermin, N.D. (2005) **It's About Time: Understanding Einstein's Relativity.** Princeton, NJ: Princeton University Press.

Migkas, K., Schellenberger, G., Reiprich, T.H. Pacaud, F., Ramos-Ceja, M.E. and Lovisari, L. (2020) Probing cosmic isotropy with a new X-ray galaxy cluster sample through the $L_X–T$ scaling relation. **Astronomy and Astrophysics**, 636: A15 (April 2020) - https://www.aanda.org/articles/aa/pdf/2020/04/aa36602-19.pdf [Accessed 6th May 2021]

Pathria, R.K. (1972) The universe as a black hole. **Nature**, 240: 298–299.

Penrose, R. (1965). Gravitational Collapse and Space-Time Singularities. **Physical Review Letters**, 14 (3): 57–59.

Penrose, R. (1979) Singularities and Time-Asymmetry. In Hawking, S.W. and Israel, W. (eds.). **General Relativity: An Einstein Centenary Survey**. Cambridge: Cambridge University Press. pp. 581–638.

Penrose, R. (2004) **The Road to Reality: A Complete Guide to the Laws of the Universe.** London: Vintage.

Penrose, R. (2012) **Cycles of Time: An Extraordinary New View of the Universe.** London: Vintage.

Poplawski, N.J. (2010a) Nonsingular Dirac particles in spacetime with torsion. **Physics Letters B**, 690 (1): 73 – 77.

Poplawski, N.J. (2010b) Radial motion into an Einstein-Rosen bridge. **Physics Letters B**, 687 (2–3): 110–113.

Poplawski, N.J. (2010c) Cosmology with torsion: An alternative to cosmic inflation. **Physics Letters B**, 694 (3): 181–185.

Poplawski, N.J. (2011) Matter-antimatter asymmetry and dark matter from torsion. **Physical Review D**, 83 (8): 084033.

Poplawski, N.J. (2012) Nonsingular, big-bounce cosmology from spinor-torsion coupling. **Physical Review D**, 85 (10): 107502.

Poplawski, N.J. (2014) The energy and momentum of the Universe. **Classical and Quantum Gravity**, 31 (6): 065005.

Poplawski, N.J. (2016) Universe in a black hole in Einstein-Cartan gravity. **Astrophysical Journal**, 832 (2): 96.

Pramoda, K.S., Saha, R., Jain, P. and Ralston, J.P. (2008) Testing Isotropy of Cosmic Microwave Background Radiation. **Monthly Notices of the Royal Astronomical Society**, 385 (4): 1718–1728.

Pramoda, K.S., Saha, R., Jain, P. and Ralston, J.P. (2009) Signals of Statistical Anisotropy in WMAP Foreground-Cleaned Maps. **Monthly Notices of the Royal Astronomical Society**, 396 (511): 511–522.

Reichenbach, H. (1958) **The Philosophy of Space and Time.** New York: Dover.

Rovelli, C. (2017) **The Order of Time.** London: Penguin.

Smolin, L. (1992) Did the universe evolve? **Classical and Quantum Gravity**. 9 (1): 173.

Smolin, L. (2019) **Einstein's Unfinished Revolution: The Search for What Lies Beyond the Quantum.** London: Penguin.

Stuckey, W.M. (1994) The observable universe inside a black hole. **American Journal of Physics,** 62(9): 788–795.

Unger, G. and Popławski, N.J. (2019) Big Bounce and closed Universe from spin and torsion. **Astrophysical Journal**, 870 (2): 78.

Vallarta, M.S. and Feynman, R.P. (1939) The Scattering of Cosmic Rays by the Stars of a Galaxy" **Physical Review, American Physical Society**, 55 (5): 506–507.

Wood, R. (2019) **Gravity-Light Energized Waves as the GLEW holding the Multiverse together: rethinking the composition and function of black holes, dark energy and dark matter.** General Science Journal, April 2019: https://www.gsjournal.net/Science-Journals/Research%20Papers/View/7755

Wood, R. (2020) **The new Big Bang Theory, Black Holes and the Multiverse explained (Gravity-Light Energized Waves as the GLEW holding the Multiverse together: rethinking the composition and function of black holes, dark energy and dark matter)** Second Edition. London: Amazon UK.

Ziefle, R.G. (2011) On the "new theory of gravitation" NTG). **Physics Essays,** 24 (2): 213-239.

Printed in Great Britain
by Amazon

COPYRIGHT

BY G.W CASSIE

SAM, MONTY, AND FRUITCAKE
THE WITCHES CAT

BY G.W CASSIE

DESCRIPTION

This is the story of the three witches cat, who turn out to be very different from the usual black cats that witches love so much.

You have Sam the Big Ginger Cat, Monty the white Cat, and Fruitcake the small and cute little cat.

Table of Contents

Page 1

Witches Flying In Stormy Weather

Whizzing at the speed of light through the dark night, the three witch's pass through the dark thunder clouds.

Shaking so much, as the wind and rain hit the three witches, Izzy's false teeth came chattering out.

Catching them she stuffs them into her pocket.
Now her gums were flapping about in the wind.

Agnes and Sybil, her sisters couldn't stop laughing.
'Mum always said, ''brush your teeth at night, Izzy,'' but no, you wouldn't listen.'

Izzy stuck her tongue out at both of them.
Before landing in their deep dark forest.

'WHERE ARE THE CATS, IZZY?' As they all turned around looking at their brooms,
Then Under their brooms.

'Drat I'm sure I secured those scraggy old cats' on properly. I put two seat belts around them this time, I'm sure I did.' as Izzy kept on looking.

Agnes and Sybil weren't convinced, looking at Izzy doubtfully.

'It's not my fault they were as old as the hills, and couldn't hang on.'

'Come on then. we better go back and look for them,' says Agnes.
'What. at this time of night? I'm hungry, very wet, and it's too dark, We'll never find black cats at night,' as Izzy put on her sad face.

We need the cats, Izzy, It's where we get some of our power from. so get on your broom and start looking.'

'*Drat*,' as she took off into the night with her two sisters.

Page 2

The Witches Sitting On a Log Getting a Moldy Brownie Thrust Into Their Noses

The three witches flew through the night, looking for their cats, only landing for a rest as the sun came up.

'What are we going to do if we can't find the cats? The other witches will start ordering us about.'

'OH, i don't like that. we're supposed to be the meanest witches around, and i quite like being mean.'

'ME TOO.'

'And Me.'

Agnes, Sybil and Izzy sat on a log with their heads in their hands, kicking around the leaves.

'Has anybody got anything to eat?' asks Izzy.

Sybil rummages threw her pockets. 'I'm sure I had some toad brownies in here somewhere.
'Here they are,' as she thrusts three green and moldy brownies at Izzy's face.

'Oh no thank you,' as she gets a waft of stale brownies. 'I'll wait till later-you have them,' as she folds her nose.

Sybil stuffed them back into her pocket. 'Suit yourself I'll keep them for later.'

Izzy and Agnes looked at her in disbelief, shaking their heads from side to side.

'Don't you think you've kept them long enough, Sybil.'

'Oh no, the greener, the better,' as Sybil licks her lips.

'Come on you two, 'get off your bums' as Agnes grabs her broom.

'SSHH, you two, I can hear cats playing in the distance.'

'LISTEN!'

All three witches listened intently

'I hear it.'

'Me too.'

Creeping as quietly as possible to the sound of cats playing.

The three witches poke their heads through the bushes.

Cats Playing In Mud

A huge smile came across their faces. Watching three young cats playing around in the mud.

'What do you think, sisters?'

'Well we need new cats- the other ones were to old- and they look like black cats,'
as Agnes squints,

'Let's get them,' says Sybil with an evil smile- With her shiny white teeth glowing.

'You can close your mouth now,' says Izzy. 'I get the message.'

Grabbing their brooms, they shoot off low to the ground towards the three young cats.

'Yikes, as the big cat spots the witches coming. 'Run you two.'

'witches, as the other two cats look around.

It was no use, as the brooms swept in low, grabbing the cats one at a time.

Page 4
The Brooms Waving Goodbye To The Cats

MEAW, MEAW, MEAW,'

The three little cats screeched all the way back to the witches' woods.

'OH, my God, my ears. I've never had such sore ears listening to three cats before.'

As Izzy and Sybil rub their ears.

.

Agnes had a confused look about her.

'WHAT ARE YOU SAYING?' as she leans forward.

'OUR EARS ARE SORE,' shouts Izzy.

'Your ears are sore.'

'YES, OUR EARS ARE SORE,' as they looked at her quite puzzled.

'My ears are ok,' while she picked out her ear plugs.

'Where did you get ear plugs from?' they both ask.

'A handsome young wizard gave them to me years ago. I always keep them in my pocket,'

'Ooooh AGNES and the WIZARD up a tree K, I, S, S, I, N, G.'

'It was years ago, you nitwits.'

'Mmm, I hope it wasn't Ralf, the wizard from rat valley- he kissed me first,' says Izzy.

'NO it wasn't Ralf,' as she tried to change the subject quickly.

The three cats stood there looking quite confused, looking at one another.
The biggest cat nods his head to the other two and takes one step back,
the other two follow.

Slowly they creep further and further away while Sibil and Izzy
Quizzed Agnes about the wizard.

The brooms on the other hand sat down watching the cats. 'do we smell, broom one asks the other two.
'we don't think so, as the other two smell under their arms.

'Why are the new cats trying to escape then, 'We should tell the witches.'

'NO don't.'
'Why not.'

'I don't like cats, 'look what their claws do, as broom three rubs the claw marks
on his bark.

'You've got a point, as the brooms waved goodbye to the cats.

Page 5
Cats Hiding Under The Cape.

'I think we're safe now,' as the cats turn around.

'*AAAARRRGGGHH,*' as the cats nearly jumped out their fur.

'Going somewhere cats?' as the witches stood in front of them.

'There's no escape, no matter what direction you take, you cats will always land back at us.'

'Well until we trust you, then we'll take the spell off.'

The three witches give the cats an evil smile, sending shivers down their tails.

'HELLO, HELLO,' a deep voice came from the edge of the forest.

'OH, no, it's MUM. Hide those cats.'

'What's with, Agnes?'
'Your capes - hurry up you two,'

Sibil and Izzy throw their caps over the cats.

Just six eyeballs staring up as the cats peeped out from under the capes.

'Hello, mum. you look lovely today
'what brings you to our lovely forest?'

As they stand in front of the cats trying to conceal them.

'Am I not allowed to visit my three beautiful daughters,' as she hobbles over, holding her walking stick.

'What are you hiding?' as she pokes her walking stick at her daughter's.

'Ouch stop!'

'OUCH! get that stick off me mum.'

'Move over, daughter,' as she pushes her way through.

'Mm, what have we here?' as she lifts one cape at a time off the cats.

Page 6

Old Mum Witch Pocking Cats With Her Walking Stick.

'Where's Terrance, Rex, and Spike?' have they died and gone to cat heaven?'

'Not exactly sure what happened to them, Mum. we did have them on the brooms until we landed,'
as Agnes and Sybil eyeball Izzy.

'Don't blame me. i'm sure i put the seat belts on, and it was very windy - i nearly lost my teeth,'

Mum witch poked at the three little cats, 'Are you sure these cats are any use? look at them.
They are very small, especially this one,' as she pokes at him.

'*OUCH,*' the little cat put his paws up to his mouth.

'What happened there?' the big cat put his paws to his mouth.

The middle cat stood there shocked.

'WE CAN SPEAK!'

'Yes you can speak. it's part of being a witches cat,' as the old witch patted the cat on the head.

'Yuck! he's got messy fur,' as she wipes her hand on her cape.

'Right, I'm away. good luck training those three little cats.
You're not getting any younger ladies.
Remember how long it took you to train Terrance,

Spike and Rex.'

'Thanks, Mum,' as they all wave her goodbye.

'That was fun, the old misery guts.'

'Come on, cats. I'll show you around our forest,' as Izzy led the way.

The cats started speaking posh to one another just for some fun.

'Hello my name is Cecil silly wiggle.'

Ooh, me next. I'm Frederick sidebottom, nice to meet you, my good men.'
Ooh me next. Hello I'm Charles Wiggly Bits,' as they all giggle.

Izzy wasn't amused, remembering how much work the last cats were when they were young. Full of life and very mischievous.

Page 7
Izzy Daydreaming About Honey.

'Settle down cats,' she shouts back.

'This is where those delicious frogs live,' as she points to the huge pond, licking her lips.

'YUCK! I hope we don't have to eat any.' the little cat mutters.

'Over here cats.' as they were falling behind again.

'Sshh, be very quiet cats.'

'This is where the honey bees live.'

'We used to love honey before those pesky bees got too clever.'

'The last time we tried to get honey,' we got chased out,'

'I miss the taste of honey,' as Izzy closes her eyes and dream's a little
'Honey on my toast,
'honey on the roast,
'honey in my soup,
'and honey on my stew.'

Izzy walks on, daydreaming.

'YUCK, squirmed the little cat. 'Honey in your soup?'

'And on their stew,' the other two squirmed.

'Come on!' Izzy shouts again, as the three cats try to keep up.

'Right try not to go in here. This is where the brooms hang out and they hate cats.'

'Why do they hate cats?' the middle cat asked.

'Would you like sharp cat claws stuck onto the side of you?'

The three cats pinged out their claws to take a look.

'GOOD point, Izzy,' they all agreed.

Making their way back to the house, the three cats get shown around the house by sybil.

'This is where you're sleeping,' as they look down to three blankets that hadn't been cleaned for years as the flies swarming around them,

Page 8
The Cats Rubbing Soot All Over Themselves To look like black cats.

'This is our potion room. Try not to touch anything in here - you might change colour or worse turn into a frog or even worse a dog.'

'Is there anything to eat? I'm rather hungry,' the little cat pipes up.

'Mmm, I've got toad brownies or porridge, which one?'

'PORRIDGE,' as the three cats flop down on their blankets, only for a puff of dust to bellow up.

'Eat your porridge and get a good night's sleep cats, you have work to do tomorrow.'

'WORK?' What kind of work?' the cat's ask.

'Wood to collect, leaves to clear away, and training on how to look mean like a proper witches cat,' as Izzy showed them her meanest face.

'yuck!' as the three cats cringe.

'Agnes,' the little cat pip's up. 'Why do witches need black cats and not beautiful colourful ones?'

'black cats can't be seen at night when we're out on our brooms scaring all the town folk.

'They're more evil - looking and they don't taste as good as colourful cats.'

The three cats' mouths dropped.

'Why do you ask?' askes Agnes.

'NO REASON,' as they all hide under their blankets.

The next morning the cats were up early.

'How are we going to keep ourselves looking like black cats, you two?' as the biggest cat looked around the potion room for ideas.

'We could rub black soot all over ourselves from last night fire,' as the middle cat rubbed it all over. 'Look, it's working.'

The three cats took to cleaning out the fire place every morning making sure they looked like black cats.

Page 9
Fruitcake Singing To The Bees

As the weeks passed, the training was going well. They were hanging onto the brooms as the witches speed through the night - even learning how to scare people was fun.

The little cat was still worried about getting eaten so he decided to sneak into the bees' forest one day.

The bees were flying all over, collecting nectar from all the colourfull flowers.

The little cat sat watching, mesmerized by how beautiful they were working together.

How do I get some honey? he thought to himself, without getting chased out the woods.

Watching for a while, he remembers his mum used to sing to him at bedtime when he was a kitten.

The little cat starts to sing…

"Red and yellow, pink and blue
 these are the flowers loved by you,
Collecting all your nectar for me and you,
To make our breakfast tasty and good,
Rest little honey bees, sleep, sleep, sleep,
I'll only take a little for my friends and me,
Rest little honey bees, sleep, sleep, sleep,
You won't even notice I've ever been"

All the bees start to fall asleep, floating down landing on flowers, even his head got covered.

'COOL,' I didn't expect that, as he opens up the hive and takes some honey out just before the bees wake up.

Running back, he pops some honey into the pot of porridge while the witches weren't looking.

'Breakfast time.' the little cat shouts.

'BREAKFAST TIME,' HE SHOUTS even louder, startling the witches.

'We make breakfast around here little cat,' shout the witches, looking at him menacingly.

'NOT today you don't,' now sit down, the little cat says sternly.

The witches sit down as they had never been told what to do before, it had confused them.

Page 10

Izzy face smiling eyes closed eating the porridge

The little cat puts a bowl of porridge down in front of them.

Izzy tries her porridge first, her eyes close and a smile comes across her face.

He's poisoned her,' shouts Agnes.
'LOOK,' I've never seen Izzys face like that before,'

Sybil panics as she had just taken a spoon full.

'*Mmmmm*, this is gorgeous,' says Izzy, eventually opening her eyes. What have you done to make this so good?'

'I got you some honey Izzy, to make you happy.' as the little cat smiles.

'I can get you more if you let me help cook.'

Sybil and Agnes gobble down their porridge.

'WOW, this is excellent little cat,' as Agnes pats him on the head.

'Give him the job Izzy, we can't make our porridge taste like this.'

'COOL,' as the little cat jumps for joy.

'Can you get me some honey,' Sybil askes with a big cheesy smile.

'What for?'

'If you get me honey, i will show you how to bake cakes,'

'OK,' as he darts away back to the beehive.

Singing his little song the bees fall asleep again.
Letting him get some more honey before the bees wake up again.

'Here you go,' giving Sybil some honey.

Sybil and the little cat spent the rest of the day baking.

First out was the fruitcake.

The little cat was entranced with the smell gobling some down.

'Mm, you like the fruitcake little cat.'

Fruitcake Entranced With The Smell Of Fruitcake

cakes into tubs to keep them fresh.

As the smell disappeared the little cat came out of his trance.

Sam and Monty were lying on their beds smiling away.

'Fruitcake,' did we hear fruitcake, as everyone else came rumbling in.

'Help yourselves, as Sybil puts it on the table.

The little cat's nose followed the smell of the fruitcake, as he stood there transfixed to the spot.

'We've came up with names for you three cats since weve decided to keep you,'

As Agnes points to the biggest one. you can be SAM, as your big and strong,

'Were calling you monty, and the little one was going to be called sid,'

'But i think we'll call him fruitcake, for now, since he likes it so much,'

The little cat wasn't listening, he was still transfixed on the cake.

Sam and Monty giggled, wait till he wakes up from this, i can't wait to tell him his new name,

After dinner, Izzy puts the

'What are you two up too?' askes the little cat.

'HELLO FRUITCAKE,'

The little cat looks behind him, 'whose fruitcake, who are you two speaking to,'

'YOU, that's your new name, FRUITCAKE,

'I CAN'T be, I wanted bruce the mighty, as he flexes his little muscles.

The witches were laughing so much Izzy's false teeth came out again making the witches laugh even more.

Fruitcake scowled at them, 'i want to be called bruce,'

Izzy gives the little cat a hug.

'Listen I will put a pencil mark on this door frame if you grow higher than this mark we'll call you bruce the mighty, ok.'

'OK, as little fruitcake shrugs his shoulders, looking at his brothers angrily.'

Page 12

Everyone On Brooms Racing Back Home To Miss The Rain

The next day, the three cats got to go to the Sunday market so Sybil could
sell all her cakes and scones.

Mingling with all the other witches and their cat's. looking at all the lovely food for sale.

From fly jam, to frog marmalade, and all makes of cakes

There was fly and mite muffins,

Frog and toad cake.

Nettle and slug soup.

'Look fruitcake, your favourite, spider and fly fruitcake,' points Agnes.

'There were spiders, and fly's, in the fruitcake?' as the cats turned green.

'Yes, how else would you make it,' askes Izzy.

'With no spiders or fly's the next time,' we hope.

After Sybil sold all her cakes it was home time.

'Come on cats,' shout's Sybil 'home time.'

'We better hurry Agnes, look the rains coming our way and i hate frizzy hair,'

'COME on then,'

'*CATS*, shrieks Izzie get on the brooms, or we'll be soaked,

'OH NO, come on you two, we better get home before the rain gets us.' says Sam.

The witches raced back to the forest with the rain hot on their tails.

'Come on Izzy open the door,'

Page 13

The Mud And Soot Being Washed Of The Cats As The Rain Shoots Past

Izzy fumbled about in her pockets looking for the door keys.

'I CAN'T FIND THE KEYS?'

'I CAN'T FIND THE KEYS?' as she turns to look at Sybil and Agnes
as the rain poured down over them.

The three witches huddled together at the door, soaked to the bone, turning around looking at the cats, as a river of black soot went by.

The rain had washed all the old mud and soot off leaving a big ginger Sam, he had black stripe's and big beautiful yellow eyes.

Monty was as white as snow with a black tail, with big blue eyes,

Fruitcake was brown with black spots with a black tuft of hair on top, with big green eyes.

Smiling nervously as the witches looked them up and down.

'Izzy,' have you got those glasses we all share,' Askes Agnes.

'MM you cats look rather tasty,' licking her lips. As Izzy puts on the glasses.

'OH NO,' these are our rain jackets, Sam tells them.

As the three of them do a twirl.

'Look how cozy we look.' as the cats puff their fur up.

'Nice try wise guys, get inside,' as Izzy finally finds the keys.

'What are we going to do sisters?' as they all sat down taking turns with the glasses.

'EAT THEM! 'says Izzy with her eyes wide open and her mouthwatering.

Sam Monty and fruitcake hid under their bed blankets, just their eyeballs sticking out.

'We can't eat them you daft witch, we've put too much work into them,'

'O that's a shame,' as she looks at them with her mouth still drooling.

'We could cover ourselves in soot, before we go out the woods, say's Monty
We've managed to hide our coloures from you three so far,

Page 14
Izzy Licking Her Lips Daydreaming Of The Cats.

The other two cats smile nodding their heads.

'And you're getting hunny,' shouts fruitcake.

'Mm ok but you can't let anybody see you outside the woods looking like that, especially mum OK, as Agnes points at them.

'OK,' as they stood to attention with their paws up to their brows.

Monty wiped his brow, phew i thought we were witches stew for a while,'

Izzy was still licking her lips.

'Izzy snap out of it, shouts Agnes 'anyway the best you can eat is soup or ice cream with those gums,'

'I can suck their bones,' she snaps back.

Making the cats feel a little uneasy, before drying themselves at the fireplace, whilst watching Izzy's every move

Page 15

Sam To Heavy For The Broom.

As the weeks passed Sam grew to be a huge cat, Agnes's broom struggled to get off the ground.

'Drat broom, what's wrong with you.'

'You're cats to heavy Agnes,' look at him, are you sure he's not a tiger?'
as Agnes's broom snaps back.

The three witches got off their brooms to go think for a while.

'He is very big Agnes he maybe is a tiger,'

'He's not a tiger he just got big muscles,'

'Well what are we going to do?' we need the cats with us most of the time, says Agnes.

'They could get their own brooms,' Izzy say's hesitantly.

Page 16
The Cats Looking For The Start Button.

'In all our 332 years we've never allowed cats to have their own brooms,'

'Well we might have to start, as the brooms can't take off with Sam on the back,'

As the three witches sit down to think for a while. Sybil comes up with an idea.

'Let them build their own ones from the magic forest, that way they will have total respect from their own brooms,'

Agreed,

'Yes, says Agnes.

'Me too, say's Izzy.

'Let's get them started,' pips up Sybil.

'Right you three, you can get your own brooms, you go to the magic forest the one we don't let you in, go to the middle, and you will see the magic sycamore tree sparkling away,

'Each of you cut off a good straight branch then bind some twigs around the end then say the magic words,'

Agnes whispers the magic words to Sam.

'Off you go,'

'What right now, fruitcake pips up.

'Yes now.' as Agnes shoos him away.

'But I haven't' had dinner yet, look at me, I'm wasting away.'
as he Holds his little podgy belly.

'When you come back I'll have a big slice of fruitcake waiting for you.'

'The proper fruitcake, not that spider and fly one I hope.'

'Yes the one you like.'

The little cat was off like a shot.

'Come on slow coaches.' as he shoots past Sam and Monty.

'COMING,' as Sam and Monty grab a saw and some rope.

Reaching the magic tree, the cats climb up, Sam cuts his branch off first then Monty, followed by fruitcake.

Tying up branches at the end, the three of them sit on top of their brooms.
'What happens now Sam,' what were the magic words, as they bounce up and down.

Sam scratches his head for a while.

'*ABRACADABRA*,' NO.

'Mm,' *OPEN SESAME*,' NO.

Fruitcake got bored, and started looking for a start button.

Monty was busy tidying up his branches.

'*FRUITY TOOTY*,' as Sam remembers the magic words, the brooms shot away.

'*AAAARRRGGGHH*,' as Monty was holding on for dear life at the end of the broom.

Page 17

Fruitcake And Monty In The Pond With The Bees Hovering Over Head

Fruitcake was still upside down, holding on for dear life.

Sam had the biggest smile ever, all the way back to the witches' woods with fruitcake and Monty screaming for help.

Monty and fruitcake struggled holding on. all the time the brooms had evil smiles.

Sam came to a stop beside the witches, that was AWESOME, he shouts, watching the other two swirl around in circles.

Fruitcake landed in the pond, while Monty landed at the beehives.

Both Monty and the broom ran out with hundreds of bees chasing them, jumping into the pond beside Fruitcake to get away from all the angry bees.

Fruitcake and Monty waiting for the bees to disappear before getting out.

Page 18

Fruitcake In The Pot Of Soup

The witches were busy laughing, Practice makes perfection,' shouts Agnes as Monty and fruitcake come over to the log fire to dry up.

Fruitcake gives her one of his evil stare's.

'Can I have some cake now before I faint.' as fruitcake rubs his belly.

'Go on then but after dinner you get back on your broom to practice OK.'

'OK, if I must.'

As the cats practiced again, Fruitcake kept sliding underneath.

'Aaa,' I don't like flying like this, all my blood rushing to my brain.'

Sam and Monty crashed into bushes laughing too much with Fruitcakes saying he had a brain.

Monty grabs some rope tying on foot pegs and a rope to steer with.

fruitcake blast's away again this time with more confidence only for Sam to come shooting past putting Fruitcake into a dive, hurling towards the big pot of vegetable soup Izzy was making in the big cauldron.

SPLOOSH! as vegetables went everywhere, only for a soggy cat to peer over the side,
With carrots for earrings, kail for hair and an onion necklace.

'FINALLY CAT SOUP,' as Izzy licked her lips.

'Not a chance, as fruitcake climbed out the pot you're not getting to gobble me up, I'm far too cute.

'SAM,' fruitcake shouts up 'WATCH where you're going you big ugly tiger.'

'SORRY, shouts Sam.

The Witches Were Scaring The Villagers

After weeks and weeks of practicing, flying there brooms Sam Monty and Fruitcake
gained the respect of their brooms.

The witches were delighted, as it meant they could go back out scaring the villages around
them.

They Dive bombed the farmers, scared the milkman and made the newspaper boy pee his pants all on their first night.

'HOME time you lot we've had a good night,' Agnes shouts.

Sam, Monty and fruitcake have a big yawn falling fast asleep as soon as they get in the door.

As the cats woke up the witches were busy packing a holiday bag.

'Where are we going,' Monty askes rubbing his eyes.

'Me Sibil and Izzy are off to the witches' yearly meeting.'

'The meanest nastyest witches, that's Me, Sibil and Izzy get to demonstrate how to be
Mean and scary to all the up and coming young witches.'

Then we have to defend our title against other witches that challenge us,
It's usually Greta and her gang,

'You three are going to Mums for the week.'

'She's going to show you three some spells you can use.'

So you'll have to be on your best behaviour, and remember to rub soot on every day mum hates colourful shiny cats.'

'A WHOLE WEEK,' the three cats squirmed.

'Don't worry. as the witches took the cats over to mums 'she's as soft as a toilet brush once you get to know her.'

'If you're lucky she'll let you meet dad.'

The witches fly away leaving the three cats at Mable's front door.

The cats knocked for ages.

Two Eyeballs Looking Down A Dark Hall At Them.

Fruitcake cupped his hands to peer into the windows.

'It's no use Sam, there's no one in, it's as dark as a cave in there.'

'Look there is another house up the road, let's try that one.'

Sam Monty and Fruitcake walked to the next house, up in the woods.

KNOCK.   KNOCK. KNOCK.

Sam knocked so hard the door creaked open letting the cats peer down a dark musty hallway full of cobwebs.

Only for two eyeballs to peer out the darkness.

(To be continued)

Printed in Great Britain
by Amazon